DESPITE IT ALL

Also by Fred Pearce from Granta Books:

When the Rivers Run Dry:
The Global Water Crisis and How to Solve It

Fallout: A Journey Through the Nuclear Age,
from the Atom Bomb to Radioactive Waste

A Trillion Trees: How We Can Reforest Our World

DESPITE IT ALL

A Handbook for Climate Hopefuls

FRED PEARCE

GRANTA

Granta Publications, 12 Addison Avenue, London W11 4QR

First published in Great Britain by Granta Books, 2026

Copyright © 2026 Fred Pearce

Fred Pearce has asserted his moral right under the Copyright, Designs and Patents Act, 1988, to be identified as the author of this work.

All rights reserved. This book is copyright material and must not be copied, reproduced, transferred, distributed, leased, licensed or publicly performed or used in any way except as specifically permitted in writing by the publisher, as allowed under the terms and conditions under which it was purchased or as strictly permitted by applicable copyright law. Any unauthorized distribution or use of this text may be a direct infringement of the author's and publisher's rights, and those responsible may be liable in law accordingly. Please note that no part of this book may be used or reproduced in any manner for the purpose of training artificial intelligence technologies or systems.

A CIP catalogue record for this book is available from the British Library.

1 3 5 7 9 10 8 6 4 2

ISBN 978 1 80351 362 1 (hardback)
ISBN 978 1 80351 363 8 (ebook)

Typeset by Iram Allam in Bembo Std

Printed and bound by CPI Group (UK) Ltd, Croydon, CR0 4YY

The manufacturer's authorized representative in the EU for product safety is Authorised Rep Compliance Ltd, 71 Lower Baggot Street, Dublin D02 P593, Ireland.
www.arccompliance.com

www.granta.com

For Peggy and Hazel

Contents

	Introduction	1
1	NATURE IS FINDING A WAY	9
2	THE POPULATION BOMB IS BEING DEFUSED	27
3	PEAK STUFF IS ON THE HORIZON	37
4	TECH FIXES CAN WORK	47
5	ANCIENT WISDOM IS A SHINING LIGHT	70
6	ECO-RESTORATION IS HAPPENING	89
7	THE MIRACLE OF THE COMMONS	109
	CONCLUSION	126
	Acknowledgements	130
	Notes and Further Reading	133
	Index	147

Introduction

It is easy to be defeatist about the fate of our planet. There is a climate crisis, species extinctions are in overdrive, tropical forests are disappearing, grasslands are ploughed up, almost all our rivers are barricaded by dams, ice is melting, sea levels are rising, and pollution is choking cities and creating dead zones in the oceans. A recent international study led by Swedish scientist Johan Rockström found we have already exceeded six of nine planetary boundaries – the ecological red lines beyond which key life-support systems become increasingly unstable.

And then there is the politics. I began writing this book in the days after Donald Trump was re-elected U.S. President, following a campaign in which he said defenders of nature and even climate scientists were part of 'the enemy within'. On taking office, he followed his words with actions: leaving the Paris climate agreement, ripping up aid programmes, and firing thousands of scientists at federal agencies in charge of forests, wildlife, the atmosphere, oceans, climate policy and pollution control, as well as removing mention of climate change on many government websites, and deleting scientific findings that he and his advisers found uncongenial. This bout of denialism came weeks after a record-breaking hurricane ripped through part of North Carolina, killing dozens of people and uprooting and flattening millions of trees, and just days after unprecedented wildfires engulfed suburbs of Los Angeles.

We are, many scientists agree, in a new geological era: the Anthropocene. Humans are the dominant force shaping our planet's atmosphere, oceans, rivers and soils. Sadly, things are not going well. We need to do some urgent fixing, or our time on the planet may prove all too transient. The Anthropocene could be over almost before it begins.

So, is there hope? Can we make good on our crowded planet? Can we, despite it all, conjure up a Good Anthropocene? Most environmentalists almost reflexively think not. They see the Anthropocene as a dystopian era, with the climate out of control, nature in a nosedive to obliteration, and ourselves likely to follow soon after. They are not alone in their pessimism: many scientists have long held a dismal view of the future. Over two decades ago, British Astronomer Royal Martin Rees gave humanity only a fifty-fifty chance of surviving the twenty-first century.

Well, I agree that oblivion is one possible outcome. There is plenty to be scared about; plenty that feels unstoppable. But I refuse to be too despondent. We have fixed stuff before. You might say that fixing stuff defines us as a species. As ecologist Ruth DeFries of Columbia University puts it: 'Humanity thrives in the face of natural crisis . . . ingenuity has brought humanity back from the brink time and again.' She calls it 'the big ratchet': problems prompt solutions that create new problems that prompt new solutions.

Some of these solutions are technical fixes. In recent decades, we have banned human-made chemicals that were eating the ozone layer, curbed acid rain and urban smogs, banished ubiquitous toxic nasties such as brain-eating lead in petrol, and staved off mass hunger by making farm fertilizer from nitrogen in the air. Of course, just because we have successfully averted particular disasters in the past does not guarantee that we will or can do so in the future. But many things that appear unstoppable can be halted and reversed, if we have the will to try.

In recent decades, in spite of the political pushbacks, the climate crisis has remained high on the international agenda, and there is a concerted effort from scientists, industry and individuals to meet the challenges it presents. Indeed, many economists say that low-carbon energy generation has already become so efficient and cheap that even Trump's executive orders and punitive taxes cannot hold it back. They note that he has failed before in his anti-environmental aims. In his first term in the White House, he promised to bring back coal mining in America. Yet coal burning in the country's power stations fell during that term by 39 per cent, more than under either his predecessor Barack Obama or his successor Joe Biden. Even if he succeeds better in his King Coal strategy this time, America's greatest rival for global economic dominance, China, seems hell-bent on conquering the world with low-carbon tech. With Trump in the White House, I'd bank on it succeeding.

But tech is not our only hope. We humans have another powerful ally for remaking the planet: the living world. We often think of nature as fragile and doomed in our hands.

But all the evidence suggests that it is dynamic, resilient and unlikely to fold under our onslaught. It has survived huge asteroid hits and ice ages. And today it is bursting through pavements, regenerating forests, taking back abandoned farmland and even supercharging evolution to replace lost species. The natural world has its limits, of course, and those planetary boundaries loom large; but all is not lost.

The debate between environmental optimists and pessimists goes back a long way. For centuries, doomsters have warned that our pursuit of a better life, coupled with population growth, must end in ecological overreach and tragedy. At the end of the eighteenth century, as the Industrial Revolution accelerated, English economist Thomas Robert Malthus said mass hunger was inevitable as exponential rises in human numbers exceeded slower increases in food production. Such thinking was pernicious. It underpinned the refusal of British colonial authorities to alleviate famines happening under their watch from Ireland to India.

Malthusian pessimism stalked the twentieth century too. In a now-forgotten 1940s bestseller, American ornithologist William Vogt argued that 'three-quarters of the human race will be wiped out' by 2050. He called his book *Road to Survival*, but it was a bleak road. He proposed that 'large areas of the Earth now occupied by backward populations will have to be written off'. Two decades later, American biologist Paul Ehrlich, who had read Vogt as a teenager, wrote his own dystopian page-turner, *The Population Bomb*. It declared that 'the battle to feed humanity is over'. As early as the 1980s, 'the world will undergo vast famines. Hundreds of millions of people are going to starve to death.'

On the other side of the debate, optimists have countered with appeals to human nature and faith in scientific and economic progress. Some conjectures were a bit fanciful. Philosopher William Godwin, a contemporary of Malthus, argued that moral improvement would 'eclipse the desire for sex', and so head off demographic disaster. More practically, American agronomist Norman Borlaug, begetter of the agricultural Green Revolution in the mid-twentieth century, drew hope from the extraordinary ability of his high-yielding crops to feed our rising numbers. Meanwhile, economists such as Julian Simon of the Heritage Foundation have touted the power of free markets to solve shortages of key commodities. Yes, population growth brings more mouths to feed, he said, but it also brings 'more hands to work and more brains to think' – to come up with more solutions to emerging problems.

So far, the optimists can point to significant wins. Ehrlich's fears of mass hunger, which helped define my generation's environmental pessimism, have been proved wrong, at least so far – in large part because of Borlaug's Green Revolution. Simon famously won a bet with Ehrlich when he argued against the idea that the world was rapidly running out of resources and correctly wagered that key commodity prices would fall in the 1980s. Canadian polymath Steven Pinker documented in his 2018 book *Enlightenment Now* how famines, undernourishment, and deaths from infectious, maternal, and nutritional diseases had all plummeted in recent decades.

So far, so good. But population, incomes and consumption all continue to grow – and climate change is a truly global menace. Has the apocalypse been cancelled or merely postponed? Are we living on borrowed time, plundering the world's resources today with devastating consequences for future generations? British systems theorist Geoffrey West, in his 2021 book *Scale*, said the constant tide of innovations in which Simon placed his faith cannot ultimately keep up with the runaway pace of new threats. We are on an ever-accelerating treadmill, he concluded, and cannot hope to keep our footing for much longer.

Who is right? Can we stay on the treadmill? And better yet, can we slow it down?

Back in the 1960s, Ehrlich framed his alarmist predictions around a simple formula. Humanity's impact on the planet and its resources, he observed, is ultimately governed by three factors: the number of people the planet has to sustain, the amount of stuff consumed by each person, and the environmental impact of that stuff. He foresaw disaster because rising trends for the first two would overwhelm any improvements we make to the third.

But the evidence so far is that things are not as bad as he feared. As this book explores, we are proving ever more reluctant breeders. The first of Ehrlich's factors, rising global population, is not accelerating as he expected. Our numbers are likely to stabilize later this century and may fall thereafter. For his second factor, developed countries seem to be at or nearing 'peak stuff', and there are good grounds to think that others will follow in time. On the third factor, the gains are much more than Ehrlich imagined. The stuff we consume is being produced with a much reduced environmental impact. All this suggests that in the Anthropocene, we may be able to slow the treadmill down.

I have spent four decades as an environmental reporter, writing bad-news stories as I explore everything from local environmental disputes to Rockström's planetary boundaries. In 1989, I wrote one of the first books on climate change. I fully understand our planet's peril, and don't retract a word. But, in the time that I have been covering the climate crisis, I have seen what advances we have made, and they give me hope that we can still turn things around.

Nothing in this book is an argument for complacency. My purpose is to shine a light on solutions and offer hope in dark times. Too much pessimism can be the enemy of the very action we urgently need. And so, in that spirit, what follows are my seven reasons to be at least a little bit hopeful about our future on planet Earth. Here is how we just might be able to have a Good Anthropocene. Despite it all.

I

Nature Is Finding a Way

They are on the prowl. Forming packs in forests, sauntering through abandoned farmland, ambling down railway tracks and motorway verges, hunting even in the suburbs. Wolves are returning to Western Europe big time. Nobody reintroduced them, or prepared habitat they might like. They are arriving under their own steam, for their own reasons and in their own time. Spreading west from the forests of Russia and the Carpathian Mountains, packs of grey wolves have crossed through Germany, climbed the Alps, and headed down through Italy and across France into northern Spain. Scandinavian populations are booming too. For most of the twentieth century, wolves had disappeared from much of western Europe, remembered only in legend and nursery rhymes. Now they are present in every mainland European country, and almost certainly number more than 20,000.

These apex predators – symbols of wildness for many of us – are not alone in their dramatic return. Brown bears, jackals and lynx, too, are recolonizing Europe. If you look for it, there is plenty of good news about nature's return. Urban badlands are becoming havens for rare species, and wildlife is running riot in radioactive exclusion zones. Across the Atlantic, the felled forests of the American Appalachians are regrowing.

Admittedly, very little of the world is pristine any longer. The human hand is everywhere, and deforestation remains rampant across much of the tropics. But amid the desecration, nature is finding a way to adapt, to evolve, and to reclaim its own.

Often this resurgence goes unremarked, and the opportunity to help it along is being lost. The problem is that our idea of nature tends to mean 'unspoiled nature', a landscape or species that exists entirely apart from human interference. We revere what we regard as intact environments, while ignoring and even disdaining nature's scruffier, more feral, untamed lands that exist in conjunction with, or in recovery from, human activity.

That is understandable. Our ideas about nature have grown out of centuries of Western culture, in which the natural world has been celebrated as untainted and essentially apart from us. But a new generation of ecologists is starting to embrace, research and even love these new wildernesses that spring up alongside or in the wake of man-made destruction. Novel ecosystems, some call them. The new ecologists reject the idea that true nature only exists when it is untouched by humans or set apart from us. They also reject the idea that the primary duty of environmentalists should be to maintain that separation.

The ideal of pristine nature 'is damaging to conservation and our ability to rectify past damage,' says British ecologist Sarah Dalrymple of Liverpool John Moores University. 'It stymies action' by sidelining conservation in the huge areas of the world where nature is in recovery mode. This does not mean we should stop worrying about lost biodiversity, but we do need to stop being precious about what kinds of nature we value and celebrate. If we are able to think differently about the forms of nature we cherish, we will see that nature isn't so fragile. It fights back. In my years of reporting, I have seen plenty of reasons to rejoice at its resilience, vigour and sheer unstoppability.

Let's be under no illusion. The scale of humanity's impact on the natural world is staggering. Since the Industrial Revolution, and especially since the mid-twentieth century, we have colonized, concreted over, ploughed up, drained, dammed, and otherwise wrecked natural ecosystems at an unprecedented rate. I have seen these changes in many places.

In 1990, as a rookie reporter, I flew across the Malaysian province of Sarawak in northern Borneo. There was logging in the coastal swamps, but inland was still largely tree-covered, with occasional longhouses visible in clearings. But change was coming. Our Twin Otter plane, flying low over the forests to a one-man-operated airport in the small, isolated town of Marudi, carried people from both worlds. In the seats at the back, a Penan tribeswoman in traditional dress and with earlobes stretched to her shoulders was clutching a live chicken. But up front were suited executives, headed for a branch office of Sarawak Plywood, ready to take charge of newly established timber concessions.

I returned to Sarawak four years later to find those loggers on the rampage. They claimed they were engaged in 'sustainable' logging, and the province's chief minister, Abdul Taib Mahmud, insisted to me that 'of course we protect our forests. People here love nature.' Maybe so, but on my next visit in

2016, the trees had all gone, replaced as far as the eye could see by plantations of palm oil. I was there to report on how a defender of the land rights of local Penan longhouse residents had been assassinated.

From the oilfields of Siberia to the desert that was once the Aral Sea, and from the Amazon rainforest to the wetlands of the Brazilian Pantanal, such horrors continue. But elsewhere, especially away from the tropics, something else is going on. Nature is sometimes coming back. One reason is that, as the world's population is increasingly concentrated in urban areas, many rural regions are emptying of people. The world has reached 'peak farmland'. By some estimates, four million square kilometres – an area half the size of Australia – has been abandoned by farmers in the past two decades. In such places, the potential for nature to recover is huge.

Some of the largest expanses of rural depopulation are in the former Communist countries of Eastern Europe, where ecologist Gergana Daskalova has lived the change. She told me her story. She was just nine months old when Communism ended in Bulgaria and the state-run farm in her home village, Tyurkmen, closed. The inhabitants left en masse, including Daskalova's parents. She grew up with her grandparents in a countryside where 'houses were abandoned, fields were engulfed by vegetation, and birds like pheasants and hoopoes became a more common sight than people'.

Daskalova eventually left too, travelling abroad and forging an academic career, exploring how our world is being transformed by human activity. But what was happening back home continued to intrigue her, and she has lately returned to Tyurkmen to research how nature is recolonizing the abandoned farmland of her childhood. How, among the voracious brambles and vines, there are the colourful floral remains of old gardens, but also orchids and a profusion of wild grasses.

The story is being repeated across large parts of Bulgaria, where the 2021 census found 300 'ghost villages', mostly surrounded by huge areas of former farms that have become new wilderness. For many ecologists, such landscapes are of no consequence. They are hardly pristine ecosystems. But Daskalova sees this abandonment of farmland to nature as 'a globally important process . . . a silent driver of biodiversity change' and a 'potential source of future solutions for conservation'.

She calculates that in the past four decades, farmers have retreated from a fifth of their land in Eastern Europe – and as much as a third in Russia. This is not just a quirk of the end of Communism, Stalin's last laugh. Similar things have been happening across southern Europe too. Here the cause is as much demographic and economic as political. From Greece to Portugal, the mass exodus of young people to cities – and the ageing and eventual death of those who remain – has over the past three decades emptied villages and left behind a rich array of fields, hedgerows, roadside verges, household gardens, and pastures with a total area equivalent to the size of Switzerland. All are great ecological niches for rewilding. It's not just plant life that is thriving; wolves, lynx and bears are all on the march, helped sometimes by hunting bans and deliberate reintroductions, but sustained by new wilderness and the absence of people.

The trend is observable in Asia too. Swathes of rural Japan are depopulating. Many Japanese agricultural landscapes are ecologically rich, so there is no automatic biodiversity gain from rewilding, but the revival of nature is still a boon for much wildlife. As China's megacities grow, the country's rural areas have lost more than 200 million people. Huge areas of farmland have become unviable, especially in remote and mountainous areas where mechanization cannot easily replace lost labour. Some Chinese researchers see a vicious cycle, in which farms are abandoned and wild animals such as boar move in, damaging neighbouring farms and resulting in more land being given up. Maybe we should instead see that as a virtuous cycle for natural regeneration.

In the U.S., the Corn Belt is shrinking as the plough has disappeared from land totalling the size of Arizona across the midwestern and southern states. This change has happened in just the past thirty years, but in other places, American land abandonment is a century old. Today, the Appalachian Mountains, stretching the length of the eastern U.S. from New England to Alabama in the Deep South, are unrecognizable from their deforested state a century ago. In 2024, I went to see.

'Endless' Trees Return

Way up in the Endless Mountains of northern Pennsylvania, Nancy Baker has been watching the trees grow on land that has been in her family for 170 years. When her great-grandfather Joseph Gamble, a descendant of Irish migrants, moved in, these hills near the River Susquehanna were all forest. He cut down the hemlock and maple, beech and black cherry, and fed them into his sawmill. With the trees gone, his fiddle-playing son farmed the exposed forest floor. The soils were poor, however, and *his* son, Baker's father, gave up farming to teach in New York. Now Baker is back, taking daily walks around her family's land as her sixty-five hectares regenerate into a new forest. 'I don't sell timber,' she says. 'For me, the most important purpose is to have the forest.'

Baker's backwoods are part of a much bigger story of ecological recovery across Pennsylvania, and the length of the Appalachian Mountains. A century ago, timber was the colonists' primary resource, cut for everything from firewood to construction, and iron smelting to railroad sleepers. Lumber barons and small-timers like Nancy's great-grandfather reduced millions of hectares of forests to what a contemporary report in Pennsylvania described as 'stumps and ashes'. With the trees gone, homesteaders briefly moved in, but with poor crop yields, the land was largely abandoned, and the trees crept back. Slowly at first, and then ever more dramatically. The acorns left on the forest floor by the loggers are now fully grown oak trees, showing that even the most wrecked ecosystems can grow again.

Along the Appalachians, some twenty million hectares of forests have returned. Pennsylvania is now once again mostly forest. Yes, the sound of chainsaws is heard, and people are once again making a living from these forests. I visited sawmills and wood yards, furniture-makers, and businesses selling lumber for basketball courts and even the flooring of the Supreme Court in Washington, DC. Even so, with some loving care from landowners, nature is growing two trees for every one chopped down, and the new forests are as biologically diverse as the former ones. Most of the old species are back. They store carbon, maintain river flows and even stabilize the region's climate. A study published during my visit found that the new forests' power to cool the surrounding air is now so great that it has almost neutralized the impact of global warming across the eastern states.

This is a very democratic transformation, driven by individual landowners rather than state protection or national parks. Baker says there are more than 700,000 forest landowners like her across Pennsylvania alone. She has trained as a forest ecologist and now teaches forest stewardship to people like her. At night, at her small timber homestead in her ancestral woodland, she sleeps content in a bed carved from maple that her great-grandmother brought by oxcart from Vermont. 'It's been storing carbon for 200 years,' she says. 'That's a nice thought to wake up with.'

If such ecological restoration can happen in America, then surely it can happen anywhere. So how far can it go? Can the rebirth of Appalachian forests be replicated across the world – in the tropics as well as in temperate lands? In many places, the return of trees is already happening. Europe has a third more trees than it had in 1900, most of them from natural regeneration. According to a study by Greenpeace, forests are regrowing across an area of abandoned Russian farmland twice the size of Spain. Irina Kurganova of the Russian Academy of Sciences calls it 'the most widespread and abrupt land use change in the 20th century in the northern hemisphere'. And while deforestation continues in the tropics, the surprising truth is that the world has more trees – and more natural trees – than it did half a century ago.

Green Shoots in the Badlands

The retreat from farming is the largest driver of this accidental rewilding. But there are plenty of other adventure playgrounds for resurgent nature. Old industrial sites and their polluted soils are among the best, because they provide myriad weird and wonderful ecological niches where weird and wonderful species can get going. Britain, as the birthplace of the Industrial Revolution, has been among the earliest nations to deindustrialize. Researchers are finding that waste dumps and old chemical works, railway sidings and metal mines, abandoned sewage farms and half-forgotten junkyards are becoming biodiversity hotspots – toxic for some species, but tasty for many others.

Nature is often not where we expect it to be. Britain's top site for nightingales is a military junkyard in north Kent. Europe's largest population of great crested newts inhabits old brick pits near Peterborough in Cambridgeshire. Greenham Common, a 1980s base for American nuclear weapons and cruise missiles that was for years picketed by women peace activists, is now one of the largest heathlands in England, rich in orchids, heather and gorse that sustain varied insect populations, as well as nightjars, woodlarks and lapwings.

In June 2013, I spent an eye-opening day exploring old industrial sites along the Thames Estuary with Peter Shaw, a botanist from Roehampton University, London. Clambering over fences and sea walls, we found dozens of rare species of bees and spiders, orchids and weevils amid rubble at the abandoned Tilbury power station, in waste lagoons in Thurrock, and amid the oil terminals of Canvey Island. Shaw reckons there are more species of bees on such sites than in the whole of the regular English countryside.

Environmentalists want to clean up these derelict 'brownfield sites', and planners want to build on them. I understand why. Urban regeneration is an important task, and such sites are messy places with great development value. But the mess is the point. Many of these scrubby, overgrown landscapes harbour Cinderella ecosystems that have more going for them than almost any greenfield site.

Even radioactivity is no barrier to many species. Two of the world's biggest continuous tracts of abandoned land are around the sites of nuclear accidents: Chornobyl in northern Ukraine in 1986, and Fukushima in Japan in 2011. These exclusion zones, largely emptied of people, cover thousands of square kilometres. The lingering fallout makes the soils and water, the fish in the rivers and the berries on the bushes all hazardous for humans. However, for nature, our absence has proved to be a huge opportunity. I travelled to both in 2016.

In Japan, wild boar, macaques and raccoon dogs have been the big winners from the exodus of people around Fukushima, marauding in the empty woodlands and invading overgrown gardens and abandoned rice fields. During my tour of the exclusion zone, I met former residents allowed in on day passes to check their old homes, hoping one day to return. But they were shocked to find their properties overrun by wildlife. Standing in the zone, I could imagine how the world might look if humans suddenly disappeared.

The Chornobyl zone was not entirely empty when I visited. Besides workers coming in every day to keep the stricken reactors safe inside a giant steel sarcophagus, ecologists ran research projects, and a few hundred self-settlers had defied the authorities and scurried home over the years to live a feral existence in the ghost towns. I met one, Markeyevych Federovych, who lived only a couple of miles from the site of the disaster and harvested a resurgent nature for his kitchen table. Around that table, he served me good vodka, flavoured with radioactive local herbs, offered me berries and mushrooms from the overgrown hedgerows, and boasted how he fished in the radioactive river. After three decades of irradiation, he seemed well on it. 'We only die of old age,' he said of his fellow returnees.

That may be arguable, but what was very clear was that, radioactive though it mostly was, nature appreciated the absence of more than a scattering of humans. The ecologists told me that since the Chornobyl accident, forests had expanded by almost 50 per cent across the exclusion zone. Amid the trees, wolves, bears, wild boar, lynx and other large animals were reclaiming their ancient stomping grounds.

No one is celebrating these nuclear disasters, but we can marvel at nature's uninhibited response.

Today, researchers tell similar stories from other parts of Ukraine, where war has turned the country into a test bed for how nature handles environments too dangerous for humans. After the Russian invasion in 2022, swathes of the country's east and south were bombarded by munitions and potholed by trench warfare. Some 100,000 hectares of forests burned, and nature reserves were sequestered for ammunition dumps. Yet, with civilians gone, 'some areas of frontline forests are increasingly reminiscent of protected areas,' forest ecologist Stanislav Viter told me. The proliferation of mines was bad news for elk and other large forest animals that detonated them. But by keeping away humans, the mined areas provided welcoming habitat for many smaller mammals, invertebrates and birds.

The war has changed the aquatic environment too. The dynamiting of the giant Kakhovka hydroelectric dam on the River Dnieper caused great short-term damage, but it also reinstated natural river flows to the Black Sea. Sturgeon are now heading upstream to their old spawning grounds on the Dnieper, and the vast exposed reservoir bed is growing vegetation rapidly. When Ukrainian botanist Anna Kuzemko made a rare field trip to the former reservoir, she was shelled by Russian mortars, but rewarded by the discovery of extensive thickets of native willow trees and reed beds that will in time become 'the largest floodplain forest in Europe'. She told me that Ukraine's engineers would want to rebuild the reservoir when the war was over, but that felt to her like an ecological travesty. 'Ukraine has a chance to restore its natural and historical heritage. We must not waste it.'

Why don't we hear more about these amazing cases of nature fighting back and creating the new wild? It is partly that both journalists and environmentalists prefer to tell bad-news stories. Having been both, I know that well. But it is also because the stats are being skewed. Most studies of deforestation ignore forest regrowth, even if it happens on land that has previously been clear-felled.

This is partly accidental. Deforestation is usually mapped by comparing satellite images from one year to the next. These images are very good at showing when forests are suddenly lost to fire, disease or chainsaws, but hopeless at spotting the slower business of recovery. And nobody is on the ground to count the sprouting seedlings, or see how in time they turn into bushes and then fully grown trees. The blindness is perpetuated because neither ecologists nor loggers are particularly interested in natural regeneration. With commercially valuable species removed, foresters don't see this 'secondary growth' as valuable enough to be economically productive. Nor is it pristine enough in conservationists' eyes to merit protection.

By some estimates, these invisible regenerating forests cover an area almost the size of Russia. They will often, given time and space, eventually recover most of their former species, like the regrown forests of the Appalachians. Belatedly, their value is starting to be recognized. Tim Rayden at the Wildlife Conservation Society in New York says we should be paying much more attention to restoring these recovered forests to their full potential, as they 'could deliver rapid biodiversity and climate mitigation benefits'. And they would achieve those things more quickly, cheaply and less disruptively than could be achieved by planting trees on cleared land. 'There is so much scope for nature restoration in degraded forests [without the need to] displace farming activity,' Rayden says.

By fetishizing the apparently pristine and denigrating everything else, we are both missing that opportunity and failing to recognize nature's powers of recovery.

Embracing Invaders

Of course, when nature comes back, it is often different from what was there before. Too much may have changed, including the climate, for the earlier landscapes to be restored or replicated. Brambles and vines can easily take over, or invasive species outcompete the depleted ranks of leftover natives. In abandoned Bulgarian villages, Daskalova has found a profusion of Ailanthus, a tree originally from China. In its native landscape, it is known as the tree of heaven, but in her village it is 'relentless and nearly impossible to eradicate,' she says.

Likewise, in the southern U.S., an Asian vine called kudzu is marauding through former cotton fields. Ecological restorers first planted it in the 1930s, during the Dust Bowl era. It grew where nothing else would, helping to restore soils. Now, however, it is seen as a curse, spreading across abandoned farmland, wrecking buildings, downing power lines and strangling native trees.

But who said nature always did what we wanted? If we are serious about rewilding, then we need to accept that it will take its own course. Often such invaders are just the first stage of an ecological resurgence. They are the first movers, and their infestations are short-lived. In time, diversity usually returns – even if it's not always the same diversity as was there before.

We should recognize that. We need to be as adaptable as nature itself, and jettison our more romantic notions about biodiversity and pristine environments. So do many old-school ecologists, says Chris Thomas, an ecologist at the University of York. They too often simply ignore any bits of nature that do not meet their expectations of what 'should' be present. 'The establishment of successful species in new locations is commonly interpreted as further evidence that the Earth's system is departing from a more desirable state,' he told me. So when biodiversity assessments are conducted, 'alien' species lurking in the undergrowth are simply not counted. As outsiders, they are deemed not to belong to the ecosystem they inhabit.

Thomas says that, in Britain, official records show biodiversity in decline, even though, if we add all losses and all gains together, 'it is unequivocally true that the total number of species in Britain is increasing'. Most of the gains – the weeds, pests and alien species – are kept off the biodiversity books. This amounts to an ecological catch-22. Biodiversity can only ever decrease, as new species are seen as inimical to the local environment, and each year millions of pounds, dollars and euros are being spent to eliminate them.

Of course, it sometimes makes practical sense to eradicate invaders. I don't say we shouldn't try to keep out pests that cause short-term ecological mayhem and economic damage. But if we choose to get rid of new species, we can't then hold up our hands in horror at declining biodiversity. And besides, ecologically as much as socially, migrants can stir things up to good effect. They interact with other species – eating or being eaten, pollinating, and rotting after death to nurture new organisms. We should embrace novel ecosystems that contain both what we regard as native species and the feral, opportunistic interlopers that find their new environment congenial. In his book about weeds, British naturalist Richard Mabey celebrated what he called these 'outlaw plants' and showed how they have had a hand in shaping not only our landscape but our civilization, becoming crops, medicines and sources of beauty and creative inspiration.

It is easy to be nostalgic about past ecosystems as a golden age. But what we imagine to be the perfect ecological state – a supposedly natural order of things that often gets called 'the balance of nature' – was mostly itself a passing phase. Nature is always moving on. And more often than we realize, humans have had a hand in what we now like to regard as natural.

Environmental scientist Erle Ellis, who runs the Anthroecology Lab of the University of Maryland, argues persuasively that most of the planet has been heavily influenced by human activity for thousands of years, including places such as the Amazon that we often think of as untouched. We have, in effect, long lived on a 'used planet', and humans and nature have not historically been at odds. The most biodiverse places are often those where people have lived the longest. By constantly tending the land, past civilizations created ever-changing mosaics of habitats where more species thrived than in places with less human interference.

So, we should be careful not to let our reverence for the pristine blind us to the environmental potential of today's wastelands, backwoods, abandoned fields, scrublands and no-go areas. In a Good Anthropocene we are likely to need more of them.

Evolution Is on a Roll

There is a final reason for believing that nature is equipped for recovery from almost anything we can throw at it. The apparent chaos of the Anthropocene is turbocharging evolution itself. Every perturbation of our environment creates an opportunity for existing species to adapt or for new ones to evolve, says Chris Thomas.

I first met Thomas in 2004, when he had just published an influential paper warning that climate change could wipe out a quarter of the world's species this century. He called it 'terrifying'. It was headline news. Two decades later, he has not changed that forecast, but he has added an important rider. Despite the coming extinction holocaust, he told me, 'the Anthropocene could raise biological diversity'. His research is showing that evolution in the twenty-first century is speeding up to fill the gaps left by the extinctions – delivering new and better-adapted species. 'The seeds of recovery are already visible,' he told me. 'Genes are jumping around. New species are beginning to emerge.'

Rob Dunn, an ecologist at North Carolina State University, agrees. In his book *A Natural History of the Future*, he says that this evolutionary spurt is happening fastest in urban landscapes, where species are adapting to city life with almost as much alacrity as newly arrived humans. This urban ecological renaissance resembles the story of the finches on the Galápagos archipelago, which Charles Darwin observed to have evolved different beaks to match the food on the different islands.

In modern cities, house mice have evolved to produce saliva with a chemical make-up better able to digest the carbohydrates available in urban food; cockroaches and bedbugs have developed resistance to pesticides and even learned to avoid sugar-impregnated bait; London's pigeons have learned to commute on the Underground. We have clothes moths and carpet beetles. Such natural innovations may not always be cuddly, but they do show that nature is on the prowl in the Anthropocene.

So where is biodiversity headed? It is undeniable that most species of plants and animals are currently in numerical decline in a human-centric world. This global disturbance is favouring some generalists – so-called superspecies – that can occupy a wide range of habitats, displacing local, rare and specialist species. Locally, however, biodiversity has sometimes increased substantially, especially in the rich kaleidoscopes of gardens, parks, toxic brownfield sites, roadside verges and so on that humanity has helped create. Such places may become crucibles for a biodiversity rebound.

Understanding this reality should change how we do conservation. We should not abandon nature to its fate. Some of our favourite species will pass into history unless we act to save them, and I'd vote for action any day. But it does mean embracing rather than rejecting the processes of change. Change in the Anthropocene is good. We should be grateful that nature is finding a way.

2

The Population Bomb Is Being Defused

I arrived late at night in Mexico City, then the world's largest city. It was 1984 and the teeming metropolis was swathed in industrial fumes trapped in the thin air at an altitude of more than 2,000 metres. I was there to attend a United Nations world population conference. On the plane, I had read blood-curdling stories of what *Time* magazine called an 'overcrowded, polluted, corrupted' metropolis that 'offers the world a grim lesson'. Its population had grown fivefold in just forty years to seventeen million inhabitants, twice the size of London.

Even the airport was jammed to capacity with people awaiting arrivals. The 2 a.m. bus transporting conference delegates to their hotels had to take swift avoiding action around a burning barricade, manned by protesting inhabitants of slums that didn't appear on any street maps. *Time*'s claim that the city was 'the anteroom to an ecological Hiroshima' didn't feel so far from the reality.

In the late twentieth century, human numbers were growing rapidly. When I was born, on the penultimate day of 1951, I would have been, by my crude calculation, the 2,536,727,035th person on the planet. Our number had about

27

doubled by 1984 and was set to reach six billion by the century's end. With ninety million more people on the planet every year, fear of the population bomb was acute and prompting brutally coercive birth-control programmes around the world. Army generals were often in charge of their nations' uteruses. Aid providers such as the U.S. government were arm-twisting countries to bring down birth rates as a precondition for doling out assistance.

The Mexico City conference doubled down on such policies. Its organizers held a ceremony to inaugurate the UN's annual awards for population policymaking. One went to Qian Xinzhong, a former Chinese military doctor and major-general in the People's Liberation Army, who devised and implemented his country's one-child policy. It required all women to be fitted with IUDs after their first baby, and if any of them had the temerity to embark on a second pregnancy, they would be forced to undergo an abortion and to be sterilized. More than fourteen million abortions had been carried out in China in 1983 alone, according to figures compiled by Qian's population police. As UN secretary-general Javier Pérez de Cuéllar put it, Qian had 'marshalled the resources necessary to implement population policies on a massive scale'.

The other award winner was India's prime minister Indira Gandhi, who during a famine had prioritized food aid to those countrymen and women who agreed to be sterilized. Delegates at the conference drew up a World Population Plan to encourage other nations to take the same road. It was hard to see that the UN held much store by its stated policy that family planning practices should be 'free and responsible'. Maybe the 'responsible' bit had priority over freedom.

In the conference halls, the main opposition to this draconian approach came not from liberal-minded types – or even from women whose bodies were on the demographic front line – but from male opponents of abortion. They included delegations sent by American President Ronald Reagan, and the Vatican, home of the Catholic Church, which had a seat at the negotiating table despite a resident population of just 735 and a declared fertility rate of zero.

The environmentalists present – who were as male as the rest of the delegates – sided with the militant wing of population control. They toured the corridors and coffee lounges, arguing that saving the world from disaster required its women to give up their right to have children. Among them was American ecologist Arthur Westing, doyen of the Stockholm International Peace Research Institute, who told me later that women should need an official permit to have babies. He wasn't sure how these permits might be allocated, but suggested there could be a market in them, so the poor could sell theirs to richer folks. I heard others quietly hoping that a global pandemic – such as HIV, first isolated the year before as the cause of AIDS – could solve our population problems by wiping out half a billion Africans, the world's fastest breeders. It was nasty stuff, a kind of green fascism.

A decade on, in 1994, I was riding the bus to the next UN population conference, this time in the African megacity of Cairo. Again, tensions were high after a recent terrorist bombing of a Red Sea resort. Every one of the 400 conference buses had a security guard with an automatic weapon riding beside the driver. The atmosphere inside the conference halls was very different, however. For one thing, a third of the delegates were women, many of them newly appointed population planners in their own countries. They included the conference's top official, Pakistani gynaecologist Nafis Sadik, and the headline-grabbing actor, feminist and UN Population Fund Goodwill Ambassador Jane Fonda. I noticed that even the Iranian TV company's camera operators were female.

The Cairo conference was, by common consent, the first time that women had been the main voices at a major intergovernmental policy event. 'Women take the lead in Cairo battle of the sexes,' headlined the *Observer*. Their impact was dramatic. Coercion and population control were off the agenda. Most delegations were promoting reproductive rights and the health of women and girls. This was a new agenda, and it was working. Giving women access to contraception and asserting their right to choose their family size was achieving what coercion could not.

Birth rates almost everywhere in the world were falling. Not least in then rural, poverty-stricken Bangladesh, where fertility fell from 5.8 children per woman in 1984 to 3.9 in 1994, and would halve again to 1.9 by 2024. China's draconian one-child policy continued for a few more years, but it was already an outlier rather than a model for the rest of the world to follow. As the baby boom subsided, demographers were for the first time predicting peak population before the end of the twenty-first century – perhaps by mid-century. Sadik had a stinging rebuke for the handful of delegates still addicted to

The Population Bomb Is Being Defused 31

the old coercive policies. 'If we had paid more attention to women [earlier], we might well have been ahead of the game as far as population numbers are concerned.'

My takeaway from Cairo was that trusting people works a lot better than coercion; and that personal liberty and action to protect the environment do not have to be in opposition. That was a startlingly gratifying piece of good news for me as a journalist and a human being.

Cairo, with its reproductive rights agenda, turned out to be so successful that the UN stopped holding big population conferences. The topic was now just one element of its actions for health and development. The population awards continued to be made, but a decade later, in 2004, the recipient was Jack Caldwell, an eminent Australian demographer, who in Mexico City had been one of a handful in his profession willing to criticize the Chinese and Indian masterminds of coercion. It was a fitting about-turn.

So, are the population battles over? Not quite yet. Our number has now passed eight billion, and will likely grow by another one or two billion. But after that, we could be in for a global population decline. Since Cairo, birth rates have fallen so low in many countries that their native populations are starting to shrink, even though people are living longer.

I saw for myself the impact of this demographic decline when I visited Singapore in 2008. In the economically dynamic city-state, most young women and men have jobs that require working long hours. They are often far too busy to have children, or even to form long-term relationships. Result: a national fertility rate that has been below 1.5 for a quarter-century now and continues to decline. Maternity wards are emptying and school rolls falling. The women of Singapore are 'going to go on baby strike,' said prime minister Lee Hsien Loong in a National Day address during my visit. Only a massive influx of foreigners was keeping up numbers. Nearly half of Singapore's citizens were born abroad.

Lee wanted to persuade his fellow citizens to have more babies, but the young people I spoke to told me this wasn't on their agenda. They were contemptuous of the cash 'baby bonuses' on offer and laughed at government officials running matchmaking services and university courses teaching love and romance. Demographic decline seemed to be baked in. And so it has proved. In 2023, Singapore's fertility rate slipped to below one child per woman for the first time.

Baby Bust

What Lee called a 'birth strike' has been spreading like wildfire across the world, among rich and poor, educated and uneducated, in developed and developing nations, communist and capitalist, Muslim and Christian, autocratic and democratic. In 2024, the average woman in Britain was having 1.6 children during her lifetime. Equivalent figures were 1.5 in Germany and Russia, 1.4 in Japan and Norway, 1.3 in Spain and Italy, and 1.1 in Malta. In South Korea, the fertility rate had fallen to a staggering 0.75 babies per woman. Of the thirty-eight economically developed nations of the OECD, only Israel exceeded 2.0. Among larger developing nations, Turkey and Vietnam were at 1.9, Mexico at 1.8, Iran at 1.7 and Brazil at 1.6.

Worldwide, the average was just 2.25 children per woman, half the rate of the previous generation. With a real replacement rate – allowing for children who die before reaching the age when they can have babies – at around 2.3, that is not enough to keep up numbers in the long run. The global rate would be lower still were it not for continuing high fertility in sub-Saharan Africa. Even there, however, big families are out of fashion as people move to cities, and mechanized farming means rural families have less need of children to help in the fields. Africa's two most populous countries, Nigeria and Ethiopia, have seen declines from seven children per woman in the 1980s to approaching four today.

Forgive me for plunging into this cascade of numbers, but they tell the story of a global megatrend of huge importance for the future of humanity. And they have been confounding demographers. Initially, some people-counters thought that the ultra-low fertility figures were a miscalculation, caused by women starting families later than their mothers and grandmothers. They predicted a catch-up as women in their thirties and forties had the babies they postponed in their twenties. But as the years pass, it looks more like cancellation than postponement. Japanese fertility has been below 1.5 for thirty years now, and Germany has been in the same situation for almost the entire past fifty years. When China finally abandoned its one-child rule in 2016, the expected rebound didn't happen. Chinese women appear to have become accustomed to having only one child, even when the state suggests they have more. In 2023, India inherited China's place as the world's most populous country. But with a fertility rate that slipped below 2.0 in 2024, India is likely to follow the downward trajectory by mid-century.

With deaths exceeding births, entire countries are depopulating – among them Germany, Japan and Russia. Helped by rampant outward migration, Bulgaria's population has fallen by a third in less than forty years. China's population peaked and began falling in 2022, and could halve by 2100 according to UN projections. The number of South Koreans may halve long before that. In parts of Germany whole neighbourhoods are being demolished because there is nobody left to live in them. Japan has nine million empty homes.

Globally, our numbers will keep on rising for a while, because past high fertility has left the world with a lot of women of reproductive age. But that won't last. The planet has already reached 'peak child'. The number of children under the age of fifteen in the world has been declining since 2021. Such numbers are an abstraction for most of us in our daily lives, but the signs are all around. Kindergartens are being replaced by care homes, and cemeteries fill while delivery wards empty.

Nothing is ever certain, but after a couple of centuries of unprecedented increases in human numbers, we seem set for peak humanity by the 2080s. Unless something changes to raise fertility dramatically, that could be followed by a demographic implosion as we enter the twenty-second century.

There is much to cheer in the defusing of the population bomb. At every scale from the local to the global, it reduces the stress on the environment. But the beneficial effects for the planet of reining in population growth in low-income countries will be limited, because each individual's consumption of resources – and the contribution those individuals make to pollution threats such as climate change – is much smaller than for people in rich nations. So, it is both misleading and morally repulsive to blame big families in the poor world for global threats caused largely by smaller families in the rich world.

And the cheers are muted because the rapid reduction in family sizes means that for the first time in our species' history, we will soon have more old people than young people. Many foresee a new demographic doomsday, caused by a rising tide of old-age dependants coupled with a reducing workforce. 'Aging is the real population bomb,' the International Monetary Fund headlined a report in 2023.

To head this off, politicians around the world from Australia to Argentina are adopting natalist policies similar to those used in Singapore, aimed at boosting a desire for babies through tax incentives and cash baby bonuses. In some parts of Russia, even schoolgirls are being paid to have babies.

Mostly, as the Singapore government discovered, such policies don't work. Nor does propaganda. The law passed by the Russian parliament in 2024 banning promotion of 'the ideology of childlessness' is also likely to fall on barren ground.

Finish the Feminist Revolution

So, what next? I think we need two strategies. We need to see older people less as a burden and more as a resource – of wisdom and knowledge. (I am writing this book at the age of seventy-three, way past the conventional retirement age, so

I admit I have a stake in this!) The ageing of our species may make humanity less dynamic and innovative, but it may also make us wiser and calmer – resulting in better decisions and fewer unreasonable demands being made on the planet.

We also need to address ultra-low fertility. It is an obvious threat to societies when couples choose not to reproduce. But we can't fall back on crude natalist methods to do this. Women almost universally now aspire to a life outside the home and beyond childbirth. Rather than bribing or browbeating them into having more babies, we need to complete the feminist revolution I saw unfold in Cairo. More rights for women need to be accompanied by more widely affordable childcare, more state support and, crucially, more responsibilities for men. These measures would make it easier for women – as well as men – to combine having children with a life outside the home.

Still today, far too many men are sticking to their old ways, expecting their wives and partners to do most of the child-rearing. But as prime minister Lee told his people: 'If husbands leave everything to the wives, or the women are forced to choose between working and having babies, they are going to go on baby strike.'

The evidence backs him up. Social scientists note that more macho societies, which often in the past pressed women into producing and raising big families, are now seeing the lowest fertility rates, as women break free. In more equal societies, fertility rates are somewhat higher and more sustainable. As I said in Cairo, personal liberty and concern for the environment do not have to be in opposition.

3

Peak Stuff Is on the Horizon

As you battle traffic jams and keep an uneasy eye on the air pollution index, it may not be immediately obvious but, whisper it quietly, we are driving less. The age of the petrolhead, the drive-in movie, the passion wagon and much more automobile iconography looks to be over. A licence to drive is no longer a rite of passage to adulthood. The modern James Dean is a rebel without a car. In fact, most developed nations reached 'peak car' more than two decades ago, with fewer kilometres being driven by fewer drivers, especially in cities. Britain, Germany, France, Belgium, Italy, Spain, New Zealand, Japan, Australia, Sweden and even the U.S. all joined the road less travelled.

Demographic factors, such as smaller families and ageing, are part of the reason for this U-turn in car use. There is also the hassle factor of driving in cities. Whatever the reason, between 1990 and 2020 the proportion of journeys made by car in London fell from more than a half to less than a third. Bikes and buses, trains and trams are all picking up the slack. But the trend extends to the suburbs and beyond. Online shopping and streaming, meet-ups by smartphone and an increase in working from home have all reduced the need to travel, whether to buy stuff, socialize or do your job.

The zeitgeist has shifted too. In particular, American youth's twentieth-century love affair with the automobile seems to be over. For the first time in many decades, most 17-year-old Americans no longer have a driver's licence. Generation Z is handing over the wheel. Online ride-hailing services such as Uber are cooler. The trend is worldwide. Many Chinese no longer regard owning a car as a symbol of affluence. Local ride-hailing giant Didi Chuxing is fast becoming the new norm in Chinese cities.

As we hit peak car, we may wonder what else in our consumer society is past its prime. It now seems there may be other limits to our consumption. Limits to our greed. Or at least our greed may be starting to take different forms – less about the accumulation of material things, and more about maximizing life experiences and embracing wellbeing. Marketing analysts, who drill down into our psyches to understand our motivations for purchases, argue that in the twenty-first century we are coming to see ourselves differently. In a widely read essay on the 'extended self', Russell Belk of York University in Toronto argues that whereas prior generations defined themselves through their possessions, today we define ourselves through our experiences.

We don't brag about the car in the drive; we instead swap stories about our adventures on vacation. We are less likely to spend spare cash on glitzy consumer goods and more likely to splash out on nice restaurants, concerts, personal trainers, pets, weekend breaks, streaming services or cosmetic surgery. High streets are filling with places where you don't actually buy new stuff, such as nail bars and gyms, play zones and hairdressers, charity shops and massage parlours.

More of us want to invest in healthier and less acquisitive lives. Besides going to the gym, we are eating fewer calories and drinking less alcohol. Global beef consumption is down from eleven kilograms per person per year in the 1970s to nine kilograms today. Even Americans, who have long been among the world's top three calorie consumers, nosh less than twenty years ago, though that has yet to impact on obesity levels.

Of course, demand for more material goods is still growing fast in places where the bare essentials are often still in short supply, and I will return to that later. But it is clear that almost all modern post-industrial societies are 'dematerializing', says Lorraine Whitmarsh of Bath University in a report for the Royal Society. Welcome to the world of 'peak stuff'.

This dematerializing arises not just from changes in our behaviour as consumers, but also from a second modern phenomenon: changes in the technologies that deliver what we continue to buy. Economists have long recognized what they call the environmental Kuznets curve, named after the Russian-born Nobel-prizewinning economist Simon Kuznets, who first identified it in the 1950s. Most economies go through a similar progression. In the early stages of economic development, a dash for growth sees countries waste their still-abundant natural resources, recklessly abuse ecosystems and fill their world with pollution. But beyond a certain point – the inflection point of the curve – they start to change tack. A combination of public anger about pollution and efforts by industry to save increasingly scarce resources sees them making more with less, and reducing the most egregious pollution. Cities become less smoggy; rivers become cleaner; people start to think about saving the last forests.

In OECD countries, manufacturers in 2015 were making over 60 per cent more product from each tonne of raw materials than they did in 1990. To take one example, a typical aluminium can today contains a third less metal than it once did.

Of course, as they grow richer, many countries will try to export their problems to lower-cost producers who are at an earlier stage on the curve. Still, the progression through the curve has been accelerating thanks to the advance of widely available modern technologies – lower-income countries, such as Cambodia, Bangladesh or El Salvador, also make our stuff using less material and less energy than rich countries did in the past.

Some of the changes allowed by digital technology promise to be even more transformative, with orders-of-magnitude improvements. A single optical fibre less than a millimetre in diameter can do the work of a thousand old-style copper phone wires. One cloud-computing data centre can do the IT work of servers in hundreds of corporate headquarters – and the work of tens of thousands of pen-pushers in days gone by. And that's before the wonders of AI.

Much the same applies to the technology in our daily lives. Consider the smartphone in your pocket. Yes, making that phone requires scarce metals and rare earths. But a single phone can replace so much of the stuff that once filled our houses: daily newspapers, telephone directories, maps, cameras, alarm clocks, radios, reference books, flashlights, diaries, CD players, PCs and more. The knock-on impacts can be huge. Now that most phones have a camera, global production of stand-alone cameras has crashed by more than 90 per cent. Because we can video-call from home, many of us only occasionally commute to an actual office.

This dematerialization of our economies and lives has rather crept up on us, but the data do not lie. According to the U.K.'s Office for National Statistics, Britain's consumption of materials peaked in 2001. In the thirteen years that followed, the personal materials footprint of the average Brit – in textiles, construction materials, metals, food, fossil fuels and so on – fell by more than 30 per cent, while resource productivity rose by almost 60 per cent. That is even after allowing for the materials footprint of imported goods. British electricity demand has also fallen – by more than a quarter since the turn of the century – with industry showing the biggest fall but domestic use also declining.

Having led the world into the Industrial Revolution two centuries ago – for a while, Britain mined 80 per cent of the world's coal and most of its iron – the U.K. now seems to be leading the world towards dematerialization. According to the European Environment Agency, EU countries consumed 18 per cent less material in 2020 than they did in 2008.

Even North Americans, the inventors of modern consumerism, are making do with less. They still use something like twice as much material per head as the global average. But U.S. consumption of steel and copper, paper and plastic, aluminium and chemical fertilizers has either fallen or remained stable for three decades now, despite a rising GDP and a growing population. With less stuff to move around, its highways have seen peak truck, as well as peak car. American water consumption has been declining for four decades too, as farmers use irrigation water better, and households adopt low-flow toilets, washing machines and showers. Donald Trump's 2025 Presidential edict to end federal limits on flows from shower heads might tweak that trend, but not by much.

Weighing Our Anthropomass

I don't want to be misleading about peak stuff. We are still remaking our planet in fundamentally destructive ways. Thanks to the larger population and the growing demand from developing countries, the world now consumes more than 100 billion tonnes of materials every year, three times more than in 1970. An average citizen is responsible for the creation and use *every week* of human-made stuff equal to more than their own body weight, half of it in the form of minerals, such as sand, gravel and clay, that are used largely in construction.

The weight of all the concrete, metal products, plastic, bricks and asphalt that humanity has accumulated on the planet's surface now exceeds that of all the living matter, according to an eye-catching study by Ron Milo of the Weizmann Institute of Science in Rehovot, Israel. 'Anthropomass' now exceeds biomass. We are, at the most literal level, living in a built environment, a concrete jungle. 'Our global footprint has expanded beyond our shoe size,' Milo says.

Yes, some of this material eventually gets recycled, including most of our used copper, aluminium and steel. But more than 90 per cent of it – including complex industrial chemicals, plastics, trace materials such as rare earths, and an awful lot of construction rubble – does not. All of that litters the Earth.

Most of the planet's ecosystems have been taken over by farmers to grow our food and other materials, such as cotton for clothing and rubber. Yet we remain very wasteful of these crops. The world's food farmers produce an estimated 5,300 calories every day for every citizen on the planet. The problem is that only half of these are eaten by humans. Much is fed to livestock that we eventually consume as meat and dairy products, but it takes seven calories of animal feed to make one calorie of beef. Meanwhile, as much as a third of the food we grow is left to rot in fields or warehouses, thrown out by picky supermarkets, or scraped from your plate or fridge drawer into the bin.

It remains a sorry picture. But it could be a lot worse. The good news is that since 1970, while our consumption of materials has grown threefold, the global economy has grown almost fivefold. That means that every tonne of stuff we extract from the Earth today yields 50 per cent more than it did back then. Globally, we are well past the inflection point on the environmental Kuznets curve. In many developing countries, rising consumption still outpaces efficiency gains. But the expectation now is that there, too, peak stuff is on the horizon. In China's case, it will be before the end of the decade.

Enough for a Decent Life

A serious threat to the idea of peak stuff is the world's growing inequalities. It is true that most of the world is better off and healthier than half a century ago. Covid notwithstanding, most infectious diseases are in sharp retreat. Life expectancy is up in most of the world, as are literacy rates and many other indicators of human development. Even so, the inequalities between the haves and have-nots continue to widen, both within and between countries. More and more wealth is in the hands of fewer and fewer people. The world's wealthiest 10 per cent have caused two-thirds of the global warming since 1990.

While the world is awash with the super-rich, billions of our fellow citizens still lack the basics for a decent existence. That widening chasm threatens the political stability of our world, just as climate change threatens ecological stability. After a lifetime writing about the world's environmental perils, I have come to fear the rising tide of inequality almost as much as rising sea levels.

So, a vital question is whether we can deliver on basic needs for the world's eight billion-plus people while still fixing the climate and sustaining the planet's other life-support systems. Many environmentalists have long believed – some publicly and some privately – that fixing climate change requires sacrificing such ambitions. By contrast, optimists argue that technological advances can and are 'decoupling' human well-being from environmental destruction. However, data to test the competing claims have been lacking. So, I was fascinated by a big international study, led by India-born technologist Narasimha Rao, aimed at defining a decent life and estimating its carbon cost. In other words, to find out what delivering 'enough stuff' for all would look like for the planet.

Rao knows the social landscape well. He grew up in a middle-class family in Indian megacity Mumbai, but surrounded by slums such as the giant Dharavi, backdrop of the movie *Slumdog Millionaire*. His study drew up a shopping list for a decent life for people living in such places. It covered nutrition, shelter, living conditions, clothing, health care, air quality, education, access to information and communication services, mobility, and freedom to gather and dissent. That list goes beyond widely accepted benchmarks such as those in the UN's Sustainable Development Goals. Rao reckons achieving a satisfactory level of nutrition requires not just food, but also a refrigerator to store it safely and a cooker that does not fill the home with lung-wrecking smoke. He says a reasonable quality of life also requires adequate home heating and cooling, at least one doctor for every thousand people, as well as access to a washing machine, to a bus or motorbike for travel to school, work, the shops and relatives, and to a phone and a television or computer.

Such things might not always be lifesavers, but they are 'globally desired by an overwhelming majority of people'. This makes those not having them risk 'social disengagement and ostracism'. Yet, he estimates that only two-thirds of the planet's inhabitants have even half of his ten basic requirements. In sub-Saharan Africa, more than 60 per cent miss out on eight of them. So, can decent living standards – enough stuff, if you will – be achieved without requiring so much energy that it crashes the climate?

Rao's answer is yes, it can. The basics of a decent life are almost all 'cheap in terms of energy,' he says. Around 20 per cent extra would do it, even without relying on a high-tech future. This percentage, he points out, is 'a small share of projected energy growth' in the coming years. If the rich can embrace peak stuff, then the poor can achieve enough stuff – and the planet can survive. I call that a recipe for a Good Anthropocene: one that embraces modern technology while curbing the appetites of the super-rich.

4

Tech Fixes Can Work

A 'technical fix' is often a term of disdain. It suggests that whatever solution to an environmental problem has been put on the table, it is a dodge, a failure to take more fundamental measures, such as changing government policy, consumption patterns or personal habits. I don't buy that disdain. Short of returning to hunter-gatherer societies with human numbers measured in millions rather than billions, we are in a world where technical fixes of all kinds can make our lives better and our impacts on the planet less dangerous. Not all tech is good, but we badly need the good stuff.

No issue needs technical fixes more than climate change. The problem is simple. Greenhouse gases, such as carbon dioxide, are emitted by burning fossil fuels to power our modern energy-intensive world. These gases accumulate in the atmosphere, where they trap heat, raising temperatures, triggering extreme weather and potentially pushing the world beyond tipping points from which there is no return. The good news is that the solutions are relatively simple too. We need to be generating fewer atmosphere-heating gases, and those that we do produce need to be kept out of the atmosphere. So how are the low-carbon fixes coming along?

We may despair at the glacial pace of international negotiations to cap emissions. I certainly do. I have been reporting on these talks for over three decades. I was at the 1992 Earth Summit in Rio de Janeiro, which agreed the Climate Change Convention, a promise to prevent future 'dangerous' climate change. I was at a conference in Germany in 1995, where a young Angela Merkel gavelled through a Berlin Mandate that two years later delivered the convention's Kyoto Protocol. This was the precursor to the Paris Agreement of 2015, which remains the touchstone of current climate policy.

Thirty-three years on, Merkel is long since retired, but the promise made in Rio still remains unmet. As the climate crisis becomes ever more urgent, the political will to act dissipates. Each climate COP (UN jargon for Conference of the Parties) seems to be a cop-out. In late 2024, during another episode of COP stonewalling in oil-rich Azerbaijan, a clutch of senior climate scientists and policymakers – headed by former UN climate chief Christiana Figueres, climate scientist Johan Rockström and one-time UN secretary-general Ban Ki-moon – declared that the negotiating system was 'no longer fit for purpose'. It had become a coalition of the unwilling.

Meanwhile, it is as if extreme and weird weather has been normalized, with 2024 the first year to exceed the 1.5-degree warming 'cap' set in Paris. It is extremely unlikely to be the last. Next stop: two degrees and more.

But thankfully, these interminable negotiations may not be so critical any more. The advance of greener low-emissions technology is outpacing the politics. Even naysayers such as Donald Trump – who for a second time pulled the U.S. out of the COP process in 2025 – are unlikely to be able to hold back the tide. We may lack political fixes, but we do have technical fixes that are not just blueprints but are already delivering real-world results. It is too early to declare victory. Far too early. However, carbon dioxide emissions do seem to have stabilized, at least.

For old stagers like me, the pace of technical transformation now under way has been remarkable. Back in 1992, the world's wind power was represented by a handful of tiny turbines on a hill in California; solar panels were impossibly expensive devices developed for space travel; and nobody imagined the rise of electric cars. Now, from the North Sea to the Gobi Desert, and from Texas to Tamil Nadu, wind and solar power are starting to take over electricity generation. In China, most cars driving off sales forecourts are electric, and the majority of the country's electricity generation capacity is now from low-carbon sources. In 2025, for the first time, the carbon dioxide emissions of the world's biggest national emitter fell even as its demand for power rose.

We know that carbon dioxide emitted into the atmosphere stays there for hundreds of years, so stabilizing the level of emissions is not enough. We need to end emissions by reaching 'net zero' by mid-century or the climate will spiral further out of control, but at least we know how to do it without breaking anyone's budget. Renewable energy is now usually cheaper than coal-fired power. Any extra costs are largely tied up in rewiring grid systems to tap into the new geography of generation. The only questions now are how quickly will we make this switch to renewables? And how much damage to the planet and our societies will we have inflicted in the meantime?

Yes, backsliders can cause a lot of problems, and Trump is very far from being harmless, but he is fighting a strong headwind. In 2024, 96 per cent of new electricity-generating capacity installed in the U.S. was for wind and solar power. I agree with British Energy Secretary Ed Miliband, who in 2025 said that the transition to cleaner energy is now 'unstoppable'. Indeed, it appears to be driving economic activity, rather than holding it back. The Confederation of British Industry (CBI) found that the country's 'net-zero economy' grew by 10 per cent in 2024, more than three times the rate of the rest of the economy. 'It is clear, you cannot have growth without green' were the words not of Friends of the Earth but of the CBI's chief economist, Louise Hellem.

A flight up the east coast of Britain, from London's City Airport to Edinburgh, was an eye-opener. I have written a lot about how Britain became a pioneer of offshore wind farms – partly because it is nice and windy out there, and partly because opposition to land-based wind turbines in people's backyards has been driving them offshore. Even so, the reality of seeing Britain's great new energy source spread out beneath me, the waves catching their bases and the winds turning their blades in the bright sun, was still breathtaking. From the Thames estuary, past the Humber and Tyne to the Firth of Forth, the waters have been transformed. An hour in the air took me past fifteen giant wind farms containing some 1,200 turbines, each up to 200 metres high, with a combined capacity of some 4,000 megawatts. At the time of writing, they are set to be joined by four even bigger generation hubs on Dogger Bank, a large sandbank off north-east England until now most famous for its shipwrecks.

The technology is spectacular. These turbines have fibreglass blades longer than football pitches, and computerized control systems that can maximize the power generated, whatever the wind speed or direction. Next up are floating turbine platforms that capture the power of stronger winds further out to sea, where the bed is too deep for anchored structures, and monitoring systems that detect approaching flocks of birds and slow the blades while they pass.

Most of the twenty-five biggest offshore wind farms in operation around the world are in the North Sea, powering mainland Europe as well as Britain. This is how the densely populated continent is finding room for renewable energy generation. Thanks to those turbines, Britain in 2024 shut its last coal-fired power station and, for the first time, got the majority of its electricity from renewable sources.

The rest of the world is following fast. Old coal-fired power plants are still churning out carbon dioxide and, shockingly, some are still being built, but their role is quickly declining. A staggering 92 per cent of all electricity-generating capacity brought online worldwide in 2024 tapped renewable energy – three-quarters of it solar and most of the rest wind. Rapid improvements in giant battery systems mean that the excess energy generated on sunny or windy days can now be stored for dark and windless nights. In 2024, more than 40 per cent of the world's electricity was generated by low-carbon technologies, if you include hydroelectricity and nuclear power as well as solar and wind. We still weren't on track to meet a goal agreed the year before to triple renewables capacity by 2030. But we weren't so far off.

Political bluster masks some of this progress. In the climate conference halls, the Indian government trenchantly defends its right to burn coal to fuel its industrial development. 'How can anyone expect that developing countries make promises about phasing out coal [when they] still have to deal with … poverty reduction?' India's environment minister, Bhupender Yadav, told the Glasgow COP in 2021. But in the real world, India is installing much more solar capacity than coal, because it is cheaper and quicker to build. By 2025, half of its installed electricity-generating capacity was from non-fossil fuels.

China also plays hardball at the COPs, but has now become the biggest driver of the worldwide renewables revolution. A decade ago, its mass-production methods dramatically drove down the cost of solar panels. Today, it is doing the same for batteries. Moreover, it is making the most of the technology it has developed. As of mid-2024, half of the world's solar power capacity was soaking up the sun across China. The Kubuqi Desert in Inner Mongolia was once known as a 'sea of death' that unleashed sand storms to blanket Beijing downwind to the east. Now the dunes are being covered with a 'solar Great Wall' 400 kilometres long and with enough capacity to power Beijing. By happy chance, the panels are also reducing the speed of the winds that once whipped up the sand. They also provide shade beneath which grazing pastures and farm crops can grow.

Finding Space for Renewables

That dual use of the land is a big plus, because the next major challenge for renewables is to find room for them without destroying valuable farmland and fishing waters, grabbing Indigenous territories or crowding out ecosystems. Wind and solar power require much more land (or sea) than coal mines or oil fields. By 2050, they could have a total global footprint around the size of the Czech Republic.

There is already pushback against these green land grabs. Norway's Supreme Court has ruled that the construction of Europe's largest onshore wind farm, in the far north of the country, violated the rights of semi-nomadic Sámi people to their traditional reindeer pastures. In the U.S., a federal court found that eighty-four giant wind turbines erected on the territory of the Osage Nation in Oklahoma amounted to trespass and should be removed. Fishers in both the U.S. and the U.K. have been raising concerns about how wind farms in coastal waters disrupt fish stocks and impede their activities. More conflicts will doubtless follow.

But there are ways out of this impasse, if we are willing to take them. Wind farms can go offshore, away from coastal fishing grounds, as they increasingly have in the North Sea. Unused industrial wasteland and roadside verges are abundant and often close to energy markets. Rooftops are still largely under-tapped spaces. Reservoirs are another viable location. There are already more than 600 floating solar farms on hydroelectric reservoirs across Asia. One of the biggest, with 340,000 panels, covers more than 200 hectares of the giant

Cirata reservoir on the densely populated Indonesian island of Java. The panels reduce evaporation from the reservoir as well as increase the power generated.

There are also many ways of sharing the land with agriculture. Thousands of sheep graze safely amid forests of solar panels in California. Wind farms coexist with cornfields across the Midwest. Solar panels on stilts shade crops of pumpkins by day and capture dew at night in Mexico. On vineyards in southern France, panels protect grapes from early ripening in a warmer world. China has more than 500 such agri-voltaic projects according to the World Resource Institute, incorporating crops, livestock, fish ponds, greenhouses and even tea plantations. In the delta of the Yellow River, the shade from solar panels has boosted shrimp yields by up to 50 per cent.

The term eco-voltaics has been coined to describe renewable energy systems that coexist with nature. In England's intensively farmed East Anglian fens, the Royal Society for the Protection of Birds has found that threatened bird species such as corn buntings, yellowhammers and linnets are more numerous on solar farms than among field crops. In Minnesota, the gaps between solar panels support wide expanses of meadow grasses rich in birdlife and pollinating insects. They have created a 'little prairie under the panels,' says James McCall, energy and environment analyst at the National Renewable Energy Laboratory. Replicating the Minnesota model across the U.S. could combine a solar revolution with a restoration of America's lost grassland ecosystems, he says. In a crowded world with competing demands on scarce land, surely this is a recipe to follow.

★

Renewable energy technologies have graduated with dramatic speed from backyard demonstrations to megaprojects covering hundreds of square kilometres. They are real game-changers for the global energy revolution we urgently need. But at much smaller scales, the sun and wind can provide reliable electricity to poor and remote regions of the world that remain beyond the reach of regular power grids. The key here looks like being village microgrids.

I visited Entasopia, a small, dusty community fifty kilometres down a potholed laterite road from the nearest power line, in Kenya's Rift Valley. Life here is hard. Farmers have for many years survived by irrigating a few fields to grow food, and Maasai cattle herders in their bright traditional dress come out of the bush to buy supplies. After the sun set, Entasopia used to be largely dark – until an array of twenty-four solar panels appeared in the village. They are transforming lives and livelihoods.

Peter Okoth had been struggling to make a go of his bar on the main street. Now, thanks to the microgrid, he had enough power for a big screen to show satellite sports channels and a fridge to keep the beer cold in the searing desert heat. Some evenings, seventy people showed up to watch, listen and buy. He had just bought construction materials for ten guest rooms. 'When you next come, you must stay here,' he said.

Across the main street, Margaret Mwangi had invested in a blow-dryer for the hair salon she ran in the room behind her tiny general store. Her customers' homes were filling with refrigerators and washing machines, she said. Down the road, repairman John Owino squatted in the sun outside his workshop, where he now carried out the welding that he once had to send to faraway towns. At the filling station, Nancy Kaisa had solar power to pump fuel. 'I had a diesel generator before, but this is much cheaper and easier,' she explained.

At the solar array in the backyard of the village chief, a smart meter controlled power to each of the sixty-four customers and communicated via SMS with payments software in Nairobi. Sam Duby, co-founder of the Kenyan start-up company that built and runs the project, believes such microgrids can deliver solar energy across a continent where 600 million people still have no reliable electricity. They could help deliver the UN Sustainable Development Goal of bringing electricity to all by 2030 – with a carbon footprint of virtually zero.

We can now say that renewables have the wind in their sails and the sun on their back. Thanks to them, the world's annual carbon dioxide emissions may now have peaked. Their takeover of the global energy market is probably only a matter of time. Time is an important issue, however. Every day of delay adds to climate change, which will be hard to reverse. And the history of past environmentally beneficial technical fixes suggests that, even when the evidence is clear, the potential for obfuscation by corporate players anxious to preserve their profits is immense. For them, every day of delay is a victory.

Footdraggers and Liars

Long before climate change became the main focus of environmental concern, the burning of fossil fuels has plagued our societies, our health and our ecosystems. Smogs were a curse for decades. For much of that time, polluters denied their responsibility. During the Great Smog of London of 1952, which killed some 10,000 people, a leading article in *The Times* declared that it was a 'common illusion' that coal smoke was to blame. 'Fogs ... are not parasites of coal fires and other dirt-creating human agencies that they were once accused of being,' it declared. It took another four years for the Clean Air Act to be passed, aiming to banish smogs.

China industrialized in the coming decades without learning such lessons. In 1999, I found myself wheezing in the cold, grimy streets of Shenyang, a giant coal-burning, steel-making industrial city in northeast China. It was the end of the millennium and felt like the end of the world. The stench of coal dust was overwhelming. The snow being cleared from the streets by gangs of schoolchildren was almost black. Even the cabbages piled up on street market stands had a thin smear. After a day, I was desperate to leave before my lungs gave out. I learned later that ten tonnes of sooty particulates fell onto every square kilometre of that city each week. Much of it entered the lungs of Shenyang's citizens, who were dying ten years sooner than if the air had been clean.

Yet besides the coal fumes, change was in the air. It now seems that 1999 was also the year that China's black-smoke emissions peaked. As in London in the 1950s, the health toll was finally recognized to be unbearable. Shenyang's air is still badly polluted today. But like most Chinese cities, it is less dirty and dangerous to breathe than in the recent past. Following the example of Europe and North America half a century ago, China has cut the emissions that cause smogs by 70 per cent, even as it has continued to industrialize. The workshop of the world has done this by shutting down millions of its least efficient and most polluting boilers and furnaces, and converting many of the rest to run on natural gas. Its people are living longer and better lives as a result.

And globally, we are now well past peak smog. Emissions of the particulates and sulphur dioxide from burning fossil fuels that historically have choked our city air have been in decline globally for about two decades, even as much of the world has continued to embark on rapid industrialization. The data cruncher on this, Susanne Bauer at the NASA Goddard Institute for Space Studies, calls it a 'turning point of the aerosol era'. She says the trend is set to continue as ever more countries seek to banish smogs to keep their citizens alive and healthy.

Not every country is advancing in the Chinese manner, it is true. Some cities around the world are still at early stages of the environmental Kuznets curve, and their air continues to worsen. The winter of 2024 saw record smogs across northern India and neighbouring Pakistan, with megacities Delhi and Lahore especially hard hit. There are no official records of the death toll, but hospitals were reporting millions of people seeking medical help as their lungs choked. It will be of no consolation for those who live in such places, but looked at internationally, they are now the exceptions.

★

Tech Fixes Can Work 59

One way in which Europe and North America initially alleviated mid-twentieth-century smogs was by building power stations with taller chimneys. In consequence, winds dispersed the muck that came out of them over much longer distances. Mixing with the clouds, it usually fell to the ground in rain as acid as lemon juice.

Acid rain became the scourge of ecosystems from Canada to Scandinavia and central Europe. In the early 1980s, I reported from Britain on how it was poisoning soils, blackening buildings and eating away at stone statues; from Norway on how the effluent from British coal-burning power stations was crossing the North Sea and killing millions of fish in Scandinavian rivers and lakes; and from Germany on how their own acid pollution was destroying the Black Forest and other ancient woodlands. The Germans had a word for it: *waldsterben*, meaning 'forest death'.

The acid poisoned politics too. While Germany and elsewhere reacted to the crisis with a clean-up of their emissions, British authorities, including the state-owned Central Electricity Generating Board (CEGB), for years denied the science. The diplomatic temperature became red-hot for a while, with Britain widely ostracized by the international community. The country became known for many years as 'the dirty man of Europe'.

I remember one day in 1985 interviewing Lord Walter Marshall, the rumbustious chairman of the CEGB, in his palatial London office with splendid views of St Paul's Cathedral. He denied to me point blank that his power stations could have anything to do with either dead Norwegian fish or the rather obvious erosion of the cathedral's Portland stone outside his window. It took three more years before he conceded, ordering desulphurization technology to be bolted onto his power station chimneys. By then, great damage had been done. Europe's soils, lakes and forests have taken decades to recover. The eroded stone, of course, will never come back.

What struck me then and now is that it took so long for Britain to adopt clean-up technology that had been available all along. Desulphurization had been patented in Manchester a hundred years before, but in the political heat of 1980s Britain, it was almost an official secret. I learned about it not from Marshall's large team of researchers, whose well-equipped labs I had visited several times, but from German power plant managers who were busy fitting the technology. They were amused to show me a German magazine article from 1880 describing the British breakthrough.

Such foot-dragging, driven by commerce and politics, is a recurring theme for innovative technical fixes. Electric cars were first driven on our roads in the mid-nineteenth century, long before petrol engines took over. Then, for more than sixty years, our cars ran on leaded petrol, as manufacturers systematically suppressed research showing that lead was escaping into the environment and harming our brains, and denied the existence of safer alternatives. A 2024 study concluded that the lead had been responsible for 150 million mental health disorders among Generation X Americans alone.

As for the modern adoption of wind energy, didn't the Dutch have a line in that, way back?

Another example of environmental delay was our response to the hole in the stratospheric ozone layer, gouged out by ozone-eating chemicals such as chlorofluorocarbons (CFCs) in our refrigerators and aerosols. The hole, which let through deadly ultraviolet radiation, was first spotted over Antarctica in the 1980s, by which time it was spreading and deepening so fast that by the mid-twenty-first century 'the planet would have become uninhabitable,' NASA's then-chief ozone scientist Susan Solomon said in an interview later.

The ozone layer is now well on the way to recovery. But the necessary action to ban CFCs and replace them with ozone-friendly alternatives was delayed for several years while the two main manufacturers, the U.S.'s DuPont and Britain's ICI, secretly vied for competitive advantage. Both companies denied that their products were to blame and successfully lobbied their host governments in Washington, DC and London to oppose proposals for an international ban.

Then, to some incredulity, both turned up at crunch talks in Montreal, Canada, to announce that they were now convinced by the science, and as if by magic, unveiled new manufacturing plants ready to roll out safer alternatives. Shamelessly, their host governments immediately declared their support for the ban – and the Montreal Protocol was ratified and enacted. The two chemicals giants had blindsided the world. Luckily, we survived their deceit. But, if Solomon's worst fears had been right, we might not have done.

With both acid rain and the ozone hole, we got there in the end, but they do serve as a reminder in our current travails. We need to fight off the naysayers, and act on the evidence of the climate problem before the damage is irreversible.

The Nitrogen Fix

Even when adopted in a timely manner, most technical fixes have downsides that will in turn need fixing. Arguably, aside from climate change, the most threatening of these in the twenty-first century is set to be the unintended consequence of the twentieth century's greatest technical fix: for world hunger.

For more than two centuries, we have been told that the world will soon no longer be able to feed its growing population. It hasn't happened because two linked technical fixes have kept mass hunger at bay. The first was the invention of artificial fertilizers. In the nineteenth century, farmers still sustained the fertility of their fields by adding locally available animal dung and crop waste. But supply was not meeting demand. Densely packed countries that needed to raise agricultural yields, such as Germany and Britain, began to import organic fertilizer in the form of guano – bird shit, if you will.

This idea arose after German explorer Alexander von Humboldt saw Inca farmers fetching guano from islands off Peru with large bird colonies during a visit in 1802. He brought samples home, and European chemists discovered its richness in nitrogen and other vital crop nutrients. Through the late nineteenth century, European mining companies imported guano in ever-greater quantities, first from the Peruvian islands and later from Pacific islands such as Nauru and Banaba, and parts of the Seychelles in the Indian Ocean. Whole islands were reduced to wastelands, and wars were fought for control of the precious product.

Then in 1908, German chemist Fritz Haber discovered how to manufacture chemical fertilizer from inert nitrogen gas in the air. Five years later, industrialist Carl Bosch opened the first factory, just in time to keep Germans fed during the First World War, when a British naval blockade prevented imports of guano.

Take-up of the new magic fertilizer from what became known as the Haber–Bosch process took off in the rich world, but it remained low in developing countries until a second technical fix. With fears rising that fast-growing populations in the developing world of the mid-twentieth century would bring widespread famine, crop scientists such as the American Norman Borlaug cross-bred new varieties of rice, maize and wheat that were able to absorb much more nitrogen from fertilizer than old varieties. What became known as the Green Revolution sent farm yields soaring from the 1960s. Since then, the world's population has more than doubled, but so has food production.

Today, when the global population is estimated at 8.2 billion, the world grows enough to feed ten billion or more people. India, once plagued by repeated famines, is now a rice exporter. Yes, there are still famines in some African countries, but not usually for want of food. The sad truth is that in the past half-century, they have usually happened with local granaries full but hungry people too poor to buy.

As a global average, the world needs one hectare of farmland to feed 4.3 people, compared to the pre-Green Revolution average of 1.9 people. This increased yield has brought great environmental benefits. Farmers have found it cheaper and easier to switch to the new crop varieties than to drain marshes, plough prairies or clear forests to make room for more fields. By some calculations, an area almost the size of Britain has been 'saved' from clearance for agriculture by the Green Revolution. Deforestation is now mostly to create farms to grow export crops.

But there is also a growing environmental cost. While the new crops can absorb a lot of nitrogen, they also leave a lot behind. Of the 115 million tonnes of nitrogen spread onto the world's fields in Haber–Bosch fertilizer each year, only around 25 million tonnes actually gets into the crop. The remaining 90 million tonnes ends up washed from soils into rivers. It is as ubiquitous in freshwater ecosystems as man-made carbon dioxide is in the air.

The nutrients in the water supercharge the growth of aquatic plant life, often choking rivers with algae, killing fish and making water reserves unfit to drink. When this water reaches the oceans, it creates plagues of toxic algae known as red tides. These suck up all the oxygen in the ocean waters, creating hundreds of dead zones from the Gulf of Mexico to the shores of eastern China and Europe's Baltic Sea. A few years back, a surfeit of fertilizer run-off in the Black Sea accelerated the proliferation of an alien comb jellyfish that wiped out most native species.

This nitrogen fix fed the world, but it is like a drug mainlined into the planet's ecosystems. It is a hidden driver of many of the world's species extinctions. The nitrogen emergency is one of the six of Rockström's planetary boundaries that we have already crossed. Yet most likely you have never heard of the threat. Nor have most politicians. There are no international agreements to stop it. But, in the coming decades, as we avert the climate crisis, the nitrogen crisis seems certain to come to the fore.

Scientists are already working on solutions, on technical fixes. They are finding ways to make Green Revolution plants use nitrogen much more efficiently. They are exploring techniques to capture nitrogen in field run-off and recycle it to make new fertilizer. They are even developing crops engineered to convert atmospheric nitrogen into fertilizer for their internal use – a kind of in-plant Haber–Bosch process. We have to hope that these new methods quickly prove viable at scale and that their take-up is swift.

Faustian Bargains

Back in the twentieth century, our future world was mapped out as if resources were endless. Now doing things better, cheaper, greener and more efficiently is becoming a prime focus. The technology to get the world to the right side of the environmental Kuznets curve is advancing fast, and this is where investment for innovation is increasingly concentrated.

The world's carbon-belching blast furnaces are being replaced by electric arc furnaces that recycle waste steel at much reduced environmental cost. Resource scientists predict that robotics, 3D printing and artificial intelligence could trim waste in many manufacturing industries by as much as 80 per cent. New technologies also promise to revolutionize how we make textiles and clothing. It may be possible to transform cement, whose production is currently one of the biggest sources of greenhouse gas emissions, into a benign carbon sink. One day, we may live in buildings that absorb carbon dioxide from the air.

In agriculture, the push is towards hydroponic farms that don't need soil; vertical farms that need much less land; crops bio-engineered to make their own fertilizer; computerized farming that cuts the use of water, land, fertilizer and pesticides while minimizing food lost to spoilage; and even food concocted in the lab.

To make fast progress on these fronts, governments and investors need to back pioneers pursuing ambitious ideas with risk capital and have greater tolerance for failure. Step forward California, where, more than anywhere, they believe that technology can make better lives for us all, while also cutting our environmental footprints and fixing the planet. In Silicon Valley, they call it eco-modernism.

The central message of the eco-modernists is that to achieve environmental goals, we don't have to downsize and adopt more frugal lifestyles. In fact, quite the opposite. As an eco-modernist manifesto proclaims: 'Both human prosperity and an ecologically vibrant planet are not only possible, but also inseparable.' Technology, they say, will help us make, process, use and recycle resources of all kinds far more efficiently, and cause a lot less environmental damage.

In the twentieth century, they say, our lives were transformed for the better by the automobile, the nitrogen-fertilized Green Revolution, the microcomputer and the internet. In the twenty-first century, our planet can be transformed for the better too – with renewable energy, AI and a revolution in agriculture that will free plants from soils and give land back to nature.

Some of this may be wishful thinking. Throwing clever brains at every problem won't itself deliver what the world actually needs. Like the economic dynamism of modern capitalism, technology is only a tool, a means to ends that we must decide on. Trickle-down economics hasn't delivered for the world's poor or for the environment, and in the wrong hands nor will trickle-down tech.

We also need to remember that technical fixes can be undermined by their very success. Take today's road vehicles. Recent innovations, such as more efficient engines and lighter materials, should have dramatically reduced energy consumption in cars. Only they haven't, because these gains have encouraged an epidemic of bigger and heavier SUVs that use just as much energy as the old vehicles. Likewise, more efficient LED light bulbs have filled the world with more lights. At Christmas, my street is ablaze with elaborate illuminated LED displays climbing up trees, threading through hedges and straddling rooftops. Ecologically at least, Santa Claus failed to deliver.

Economists know these perverse outcomes as the Jevons paradox, after the nineteenth-century English economist William Jevons. He noticed that James Watt's invention of a more efficient steam engine to run industrial processes didn't cut coal burning but caused it to accelerate – becoming one of the driving forces of the Industrial Revolution. Some fear much the same might arise from today's digital revolution, where cheaper, faster and more energy-efficient computing simply drives up demand. Yes, digitizing and AI should deliver big efficiency gains for industry, transportation and many other areas of activity, and those gains should reduce our ecological and climate footprints from the manufacture of material goods. But it remains unclear whether the gains will be outweighed by the dizzying energy requirements of the data centres needed. The jury is out.

Perhaps a 2025 assessment by the International Energy Agency was right to conclude that the downsides will be modest. It reckoned only a tenth of the likely growth in energy demand in the five years to 2030 will come from AI data centres. But the error bars on all this are huge. When I asked an artificial intelligence chatbot for its projection of its own future climate impact, it stressed the huge uncertainty and failed to come up with a number.

So, let's not be complacent. All technical fixes have drawbacks. Jesse Ausubel of Rockefeller University, one of our most prominent techno-optimists, concedes: 'Technology is a Faustian bargain. Any technology contains the seeds of its end,' arguing that some other superior technology 'will generally take over'. His favourite case study is New York, his home city. In the early twentieth century, the metropolis was the world's biggest, and was in danger of being overwhelmed by horses, the main means of transportation. The era's doomsayers warned that its streets would soon be knee-deep in horse manure. Then, along came motor vehicles, which got rid of the manure. Eventually, their air pollution became intolerable, resulting in exhaust filters and unleaded petrol. Soon, electric vehicles will do away with tailpipe emissions altogether. New York City lives on.

In any case, Faustian bargain or not, 'there is no turning back,' says Ausubel. 'At a billion people in the world, there might have been an alternative way of living.' That's no

Tech Fixes Can Work 69

longer possible, with half the world's population not having an acceptable quality of life. 'We have no choice but to get better at providing the services people want,' he says. 'I don't think my green colleagues have enough faith in their own scientific and technical peers.'

I rather agree. The bottom line is that technical fixes have solved big global environmental problems before. What we have often lacked is the foresight and the will to make the right choices and deploy them early enough. For climate change as much as anything else, time is running out. To have a Good Anthropocene, we need to press on, and to do so rather more wisely and speedily.

That sober message brings us back to wisdom. For want of wisdom, technological and scientific advances have done plenty of bad things over the years – from inspiring the grotesque social engineering of eugenics to making nuclear bombs. Wisdom is what we most lacked then, and need now. So where do we find it? How can we live our lives, and rule our societies, as if the essential life-support systems of our planet mattered? Where do we find wisdom to secure our future? Perhaps, I suggest, in an unexpected place: our past.

5

Ancient Wisdom Is a Shining Light

Deep in the South American bush and three days' walk from home, a tribal hunter punts down a backwater before putting aside her bow and arrow, taking out her phone and activating the GPS app. She doesn't need the device to know where she is, but she taps an icon on the screen to make a digital record of the location of the sacred grove she is passing, where her ancestors are buried. Earlier that day she had geopositioned trees that mark the border of her tribe's domain, adding to a catalogue of farms and villages, cultural sites and wildlife areas.

Her data will be uploaded later as part of a ten-year exercise by the Indigenous Wapichan people in remote southern Guyana, on the border with Brazil. The tribe is mapping more than two million hectares of its traditional forests, pastures and wetlands. The aim is to reinforce their case – so far unresolved – to obtain legal title to an area roughly the size of Wales, so they can protect it from roads, dams, mines and forestry projects planned by foreign corporations, raiders from over the border in Brazil, and their own government in faraway Georgetown.

The need is urgent. 'Our land is being taken away from us, often without us even knowing,' said Nicholas Fredericks, one-time precocious young cowboy and now tribal leader

Ancient Wisdom Is a Shining Light 71

and coordinator of the mapping project, as we ate breakfast in Shulinab village, on the banks of the River Sawariwao. 'Outsiders have a financial view of the land. They see it as money. We use the forests, but we don't destroy them.'

Besides its legal and environmental importance, the mapping project aims to reinforce tribal identity in a community where many young people leave to get education and jobs. That identity includes a connection with their land's wildlife – its jaguars, giant river otters, and endemic fish and birds like the diminutive red siskins and Rio Branco antbirds – and its spiritual places. The maps compiled on their phones include sacred creeks where they swim, sacred trees where they climb, and sacred forest groves containing their ancestral graves. Places deemed so precious that they are never shared with outsiders, whether government officials or visiting journalists like me.

The maps are backed up by interviews with elders, who describe the natural and spiritual importance of the sites. 'The elders told us how if you cut down certain trees in the forest, you will get sick and die, punished by the spirits,' said tribal official Claudine La Rose, who carried out many of the interviews. Sitting in front of her computer screen, Tessa Felix, one of the custodians of the digitized data, said she feels the ancestral connection strongly. 'When I was a child, my grandfather took me into the forest. He showed me the secret places and told me our stories. He said that I was born from the Earth, and I believe that,' she told me. 'My work now is about preserving my grandfather's legacy. Our land is like our mother. Now I want to fight for our land.'

Around the world, other tribes are similarly embracing modern technology alongside their traditions to protect their environment. Many, like the Wapichan, are digitally mapping their ancestral lands, pixel by pixel, metre by metre and tree by tree. They know their wildlife and the medicinal properties of dozens of local plants. They manage those lands using ancestral knowledge. Then they hire the best city lawyers to secure legal title and defend them from miners, farmers or their own governments.

For the Wapichan, this fight is vital to securing their inheritance and protecting the land for future generations. 'We see it as life,' Fredericks said. 'We have to win.' But the rest of the world should care too, for the tribal traditions of people such as the Wapichan are indivisible from the knowledge of how to value and protect the world's wild lands. Knowledge we all need.

Green Grabs and Blame Games

Outsiders have long had ambiguous views about Indigenous tribes and other poor rural communities with their roots in the land. On the one hand, traditional botanical knowledge has been increasingly recognized as superior. Pharmaceutical companies have made fortunes by persuading tribal elders to reveal the secrets of the medicinal properties of plants in their midst – usually with little or no recompense for the theft of their intellectual property. Most famously, the Madagascar periwinkle was long used by the island's people for a range of maladies, before knowledge of its medical properties was passed on to Europeans. Now, extracts from its leaves and flowers are employed worldwide to treat everything from malaria to diabetes and to help fight a range of cancers by restricting cell division.

Yet at the same time, outsiders have denigrated Indigenous people as ignorant ecological vandals, slashing and burning forests, overgrazing pastures, hunting wildlife to extinction and turning farm soils to desert. Back in the late twentieth century, most commentators agreed with the celebrated British environmental scientist Norman Myers that 'by far the number one factor in disruption and destruction of tropical forests is the small-scale farmer'. Media images of tropical forest destruction reinforced the stereotype by featuring belligerent-looking farmers wielding firebrands and machetes while standing in forest clearings. Such demonization was then used to justify rich environmental groups taking over those forests, banishing the inhabitants, denigrating their knowledge and fencing in the wildlife.

In the Democratic Republic of the Congo alone, anthropologists and Indigenous rights advocates estimate that 35,000 Indigenous people have been expelled from protected areas and deprived of their hunting and gathering grounds, often at gunpoint. I reported how a few environmentalists were pushing back against this approach, often called 'fortress conservation', and how it had become an ideological battle within one group, WWF. In 1993, an in-house coup there installed a new-broom director-general, Claude Martin, who publicly warned his colleagues that, in their hands, environmental protection looked 'just as narrow and selfish as the imperialism of old' and often took 'little or no account of the rights and needs of local people'.

But his staff, and national affiliates, didn't always take much notice. A decade later, as Martin prepared to retire, he commissioned me to write a short history of the organization. In it, I repeated his often-expressed view that 'too often in Africa in the 1970s and 1980s, WWF helped organize the expulsion of tribal groups from their land on the pretext of preserving wildlife. The result . . . was often to alienate the very people who had successfully shared the land with big game for centuries.' The comms team edited out those sentences from the published version.

The battle still rages. In a 2020 report, the UN Development Programme concluded that after Martin left, WWF continued to fund the brutal expulsion of Indigenous people in the Republic of the Congo who attempted to return to their ancestral lands. Two years later, the African Commission on Human and Peoples' Rights, an arm of the African Union, ruled that the Batwa hunter-gatherers should receive an apology for past expulsions from the Kahuzi-Biega National Park in the Democratic Republic of Congo, as well as compensation and permission to return home. Sadly, at the time of writing, there has been no such redress.

This matters not just for the tribes and their backers, but for the environment too. There is growing evidence of the superiority of tribal knowledge and custody of critical ecosystems. A 2014 report from the World Resources Institute (WRI) in Washington, DC, reviewing 130 studies from around the world, concluded that deforestation rates inside Indigenous reserves are typically only a tenth of those outside, and less even than in state-protected areas. 'Indigenous communities are unsung heroes of conservation,' said the WRI's land rights director Peter Veit. He was backed up by a former head of the blue-chip Center for International Forestry Research, David Kaimowitz. 'Many people still think if governments give communities the rights to forests, they will just cut them down. But the evidence shows that community control is usually the best way to conserve those forests,' he said.

The environmental gains of Indigenous management are easy to see in the Amazon rainforest, where the land rights of the 1.5 million Indigenous residents are well established in law, if not always in practice. Satellite images show how much greener their reserves are. The trees are soaking up carbon dioxide from the air as they grow, whereas the rest of the forest is releasing carbon.

The contrast is abundantly clear on the ground too, as I saw one morning while driving along a narrow track in the state of Mato Grosso on the Amazon's southern fringe. On one side of the track was cool, moist rainforest, stretching north-west for hundreds of kilometres through the almost intact Xingu Indigenous reserve. On the other side, hot, bare ground was being prepared to plant soy on a farm the size of fourteen Manhattans. According to my guide, Earth-system scientist Michael Coe, we were on the front line of deforestation in the Amazon – where the rainforest meets agribusiness, where Indigenous law meets the commercial norms of the rest of Brazil.

A century ago, the land occupied by that soy farm was still remote jungle. Eccentric British explorer Percy Fawcett had disappeared there while searching for the rumoured 'Lost City of Z'. A hundred people died trying to rescue him. In 2016, Hollywood made a movie out of the mystery. Today, more than a third of the area is deforested. The local town of Canarana, a 16-hour bus ride from Brasilia, is full of grain silos and John Deere franchises servicing big soy farms. But beyond the agribusiness frontier, the giant Xingu reserve remains 85 per cent forested and full of life – harvested, but also protected, by its 6,000 inhabitants.

Coe and I didn't have permission to enter the reserve to explore for ourselves, but its forest animals regularly strayed onto the farm. Tapir tracks and droppings were everywhere. Armadillos burrowed in the verges. Tall flightless rhea birds wandered around looking for seeds. I even saw a young jaguar saunter down the track dividing the two domains. Such sightings are a sign of what has been lost in the Amazon – but also of what remains.

For Coe, a regular visitor from the Woods Hole Research Center in Massachusetts, the boundary between farm and forest is 'the perfect laboratory' for exploring how forests interact with climate, and how the climate changes when the forest disappears. With temperatures rising, and the dry season in Mato Grosso now four weeks longer than prior to deforestation, he and local researcher Divino Silvério were keeping watch for the long-predicted tipping point – the moment when the Amazon dries out enough to trigger runaway degradation to savanna grassland. Some researchers say this could be a tipping point for the planet as a whole. If so, the Xingu may be our last line of defence.

The Real Friends of Africa's Megafauna

Ancient wisdom is protecting not only rainforests, but also a whole array of different environments, including the grasslands of sub-Saharan Africa, home to great herds of elephants and rhinos, buffalo and giraffes, zebra and wildebeest, and hunting loners like lions and leopards. We often think of this megafauna as being protected primarily within the continent's great national parks. Not so. The parks – even renowned ones, like the Serengeti and Kruger – are an increasingly small part of the story. Much of the most successful conservation today is in the hands of tribal groups, many of them cattle herders, who protect wildlife and thwart poachers in collectively owned reserves known as conservancies. Nearly a fifth of Kenya, for instance, is now under various community conservancies, and two-thirds of its wildlife live in them, sharing the land with the communities' livestock – something never allowed inside the country's national parks, whose custodians usually insist that wildlife and livestock should be kept apart.

Perhaps we shouldn't be too surprised at this success in the face of conventional conservation attitudes. After all, prior to the arrival of Europeans, Africans and their livestock had a long history of coexistence with the continent's wild animals. That is why the continent's big beasts still strut their stuff, while many elsewhere have become extinct.

I first began to appreciate this in 1998 while being shown around Kenya by its then-parks chief, the conservationist David Western. Despite his day job, he was a rarity among park managers in being an enthusiast for community conservation. He fixed a visit for me to Il Ngwesi, a newly opened independent eco-lodge owned by Maasai cattle herders, perched above a watering hole on the Laikipia plateau near Mount Kenya. Collecting me from the grass airstrip, tribesmen took me on an open-truck drive to the lodge through a tightly packed herd of about a hundred migrating elephants, many almost within touching distance. The open-air rooms looked down on an animal watering hole, where I later watched leopards by moonlight from my room, before some of my Maasai hosts left off guarding the lodge to tend their cattle on the same lands.

It was a magical experience, and when I checked it out two decades later, I found the lodge was still going strong, with ecotourism and wildlife conservation combining well with a continuation of traditional Maasai cattle raising. 'Since we started, I have seen the vegetation and the wildlife coming back,' said Kip Ole Polos, one of the founders of the lodge, who now travels the world to make the case that cattle herding is part of the conservation solution in Africa and not the problem. Local private landowners I met – many left over from colonial times – have come to agree. 'This is the future, as well as the past,' shouted Mike Dyer, owner of the nearby Borana Conservancy, as we flew in his bush plane out of Il Ngwesi and over the ancient Mukogodo Forest, an elephant-rich reserve that is similarly now under community management.

Forest Lessons in Wisconsin

There is a growing acknowledgement of the value of ancient wisdom in North America too, as I discovered on a more recent trip to Wisconsin, where I met McKaylee Duquain. Three years out of graduate school, she was in day-to-day charge of 100,000 hectares of forests that occupy almost the entire reservation of her tribe, the Menominee Indigenous community, an hour's drive from Lake Superior. This is the largest Indian territory east of the Mississippi, and foresters and conservationists alike agree that under Duquain's custody, it contains some of the best and most biodiverse hardwood forests in America.

With her boss, forest manager Ron Waukau, she decides where the tribe's loggers should cut trees, and what they should send to the tribe's sawmill. For this is no pristine forest, and the Menominee have plenty of customers for their prized lumber. 'We have been logging here since we secured our lands in a treaty 160 years ago, but we still have more trees than when we started,' Duquain told me proudly, as she looked up from the computer screen that tracked every tree, its age, size and health.

The Menominee reservation is by common consent one of the most remarkable working forests in the U.S. It contains more than a billion trees. A detailed study by outside researchers in 2018 found that, after more than a century of logging, the Menominee forest was 'more mature, with higher tree volume, higher rates of tree regeneration, more plant diversity, and fewer invasive species than nearby nontribal forest lands'.

It's not just the trees that prosper from this attention, either. The reservation is a refuge for wolves and other wildlife from the surrounding state, where, for many kilometres, almost all the trees have long since been clear-felled to make way for farmland.

Duquain says their secret lies in the tribe's well-tested skills in forestry, which combine a long tradition of common ownership of the forest with modern science and detailed on-the-ground forest inventories. Duquain's loggers adhere to tribal rules for sustainable harvesting laid down by their chief at the time the land became theirs. 'Start with the rising sun and work towards the setting sun, but take only the mature trees, the sick trees, and the trees that have fallen,' Chief Oshkosh instructed. 'When you reach the end of the reservation, turn and cut from the setting sun to the rising sun, and the trees will last forever.' His words are on a plaque in front of McKaylee's office. She sees them before she turns on her computer each morning.

Heading away from the office, I joined the tribe's chainsaw-wielding harvest administrator Mike Lohrengel, as he took a party of state and private-sector foresters on a tour to see how Chief Oshkosh's words are put into practice today. 'The forest looks pristine because we manage for the long term,' he said. 'We come into each sector every fifteen years, take out the weak trees, the sick trees, and the ones that are dying, but leave the healthy stock to grow some more and reproduce. We don't plant anything. This is all natural regeneration, and the way we do it, the forest just gets better and better.'

He looked up at the forest canopy as a flurry of snow fell. 'This is one of the most beautiful places I know,' he said. 'This forest has it all: the most species, the most diversity. These big maples and basswoods are around 150 years old. Many trees I have known individually for 30 years now. Look at this one behind us. It's got a split way up there. I'll never forget that tree till I die.' Lohrengel's party of visitors were in awe. There are no trees like this for hundreds of miles, they said.

I learned more about the Menominee forest management wisdom from its fire chief, Curtis Wayka. He was standing next to a fleet of gleaming fire engines, but he let me into a secret: the Menominee don't use them much these days. 'We spend as much time starting fires as stopping them,' he said. 'Our ancestors understood and used fire well. We are going back to that.'

Conventional European fire management practice, widely adopted in the U.S. and around the world, has insisted on zero tolerance for fire. It should at all times be suppressed. Traditional community management, by contrast, sees fires as an intrinsic part of ecosystems. Essential, even. They work with it rather than against it. In particular, most communities set fires in ground vegetation at the start of dry seasons. The aim of these 'cool' fires is to reduce the amount of dry kindling that will be available later in the season if 'hot' fires break out.

Such traditions have long been carried out in defiance of state authorities that outlaw all fire-setting in forests. Now finally, after years of being criminalized, ancient fire wisdom is gaining a hearing. Studies by Stanford University published in the wake of the 2025 fires that devastated the suburbs of Los Angeles showed that the conflagrations were substantially less intense in areas where cold burning had been carried out.

'It turns out that the educated elite were wrong, and the nominal primitives were right,' says fire researcher Stephen Pyne of Arizona State University, who first alerted me to the perverse politics of American forest fires. As the climate changes, fire managers in the West are increasingly looking to Indigenous counterparts such as Wayka for guidance on how to reduce the risk of wildfires. So now, in quiet times, Wayka told me his crews travel the U.S. sharing their expertise with state fire teams. As forest fires escalate in the U.S., their fire wisdom is likely to be in growing demand.

All Hail, Slash-and-burn

In such ways, Indigenous people almost everywhere are better able to attune their practices to their local environments, and to work in harmony with nature. Sámi reindeer herders in northern Scandinavia have a much more sophisticated knowledge than outside experts of the natural cycles of their pastures. They pass down from one generation to the next the knowledge of where and when to move their reindeer according to the seasons, the weather conditions, the growth of the lichens on which the animals depend, and the last time the land was hit by bushfires. No outsiders can match their expertise, built from centuries of experience.

Charles Peters, a forest ecologist at the New York Botanical Garden, picks up the theme. Returning from visiting the Dayak people in Borneo, he was full of enthusiasm for what he had learned from them. 'Local people know a lot more about how to manage tropical forests than we do,' he admitted. 'These guys . . . are managing 150 species in a hectare. They pay attention to every one of those species and ask how it is doing and what its requirements are. As a forester you just go: "Oh, my God, how do they do this?"'

The land cared for by Indigenous people is not just better protected, it is also often more productive. Traditional methods of farming in the tropics especially are proving superior in providing farm produce while protecting the surrounding forest. Some call this swidden farming, others shifting cultivation or, more pejoratively, 'slash-and-burn'. Whatever, despite its bad reputation among outsiders, it works.

While visiting the Wapichan in southern Guyana, I spent time with Patrick Gomes, a teacher and farmer, who took me to see his pride and joy, his swidden farm. Driving up to the forest edge along ancient trails, we passed a man with a bow and arrow on his way to distant hunting grounds, then women riding bullocks to collect forest produce. Entering the trees on foot, we crossed a stream, where Patrick pointed out a spot where he had recently found a large anaconda digesting its prey. Along the path, he pointed out wild bananas, a patch of medicinal plants, and some straight stems that he harvested to make arrows. Finally, we reached his cassava field, a small patch of cleared ground surrounded by forest.

He told me how he regularly clears plots like this, cultivating them for a year or two before moving on, to allow the soil to recover and the natural vegetation to regrow. It does so quickly, he said, pointing out a field he had abandoned the previous year, which already contained head-high new growth. 'I leave a fifteen-to-twenty-year fallow, before returning to the same place. By then it is indistinguishable from the canopy forest around.' Nature certainly liked his shifting cultivation. The forest around his clearing was alive with sound, and as we walked on to his peanut plot, two scarlet macaws rose into the sky.

Thirty years ago, Gomes would have been roundly condemned by most outsiders for being an agent of deforestation. As a 1997 review for the World Resources Institute noted: 'Shifting cultivation and the people who practice it . . . are widely perceived . . . to be primitive, backwards, unproductive, wasteful, and exploitative and destructive of the environment . . . They have been blamed for most of the world's tropical deforestation, land degradation, and climate disruption.'

Farmers like Gomes were seen as 'tradition-bound peasants, trapped by ignorance and unable to manage their resources properly,' says Thomas Tomich, a researcher in sustainable agriculture at the University of California, Davis. As recently as 2020, WWF's website declared, without qualification, that slash-and-burn fires 'don't just remove trees; they kill and displace wildlife, alter water cycles and soil fertility, and endanger the lives and livelihoods of local communities'.

But the truth is that ever since the 1990s, a growing body of scientific research has shown the ecological virtue of swidden farming. It is a 'brilliant strategy' that speeds up the recycling of nutrients that crops remove from the soil, says Ruth DeFries, an environmental biologist at Columbia University. Far from wrecking ecosystems, it can actually have a positive impact on forests, delivering increases in forest plant diversity, says Sean Downey, an ecological anthropologist at Ohio State University. Swidden farmers like Patrick have gone from environmental pariahs to environmental heroes within a generation.

Traditional hunters are also being recast as adept conservationists. Back in 2000, I visited the Indigenous Gwich'in, native Americans in north-east Alaska. Many people I spoke to said the tribe were threatening the survival of their main prey, the Porcupine herd of caribou. But the Gwich'in insisted they were the best friends of the herd. 'We are caribou people. That is how we identify ourselves; that is how we feed ourselves. We need the caribou like the Amazon Indians need the rainforest,' village head Sarah James told me.

The Porcupine herd makes one of the planet's last great mammal migrations, a 2,500-kilometre annual trek from Canadian mountains to calving pastures in Alaska's Arctic National Wildlife Refuge. That journey is facilitated by the Gwich'in, who protect habitat along the route and ensure no fences block the way. Schoolchildren have adopted caribou cows that are fitted with radio collars for the journey. In the classroom, the children give the animals names and log on to their phones for regular updates on their progress towards the pastures.

During my visit I got to see the caribou arrive in the mountains near the main Gwich'in settlement. Biologists I spoke to back then said the herd was under intense pressure from hunting, from climate change and from oil companies encroaching on its calving pastures. I wondered if the herd was doomed. Since my visit, James has become an international figure in the campaign to secure land rights for native people, in part so they can protect the herd's migration. It is a persuasive story. For under Gwich'in stewardship, the Porcupine herd has doubled in size since my visit. More than 200,000 now make the migration. They must be doing something right.

Can this symbiotic relationship between hunters and hunted survive Donald Trump's new plan to open up caribou pastures to oil drillers in the Arctic reserve? That remains to be seen. My view, however, is that the Gwich'in hunters are the caribou's best allies. And the wider lesson for conservation from the successes of ancient wisdom in ecological management is clear. As Erle Ellis of the University of Maryland, puts it: 'Empowering the environmental stewardship of Indigenous peoples and local communities will be critical to conserving biodiversity across the planet.'

Water Wisdom

Ancient wisdom has shown great sophistication in many fields, not just in ecology and the stewardship of natural resources. Traditional societies were often innovative environmental engineers too, especially when it came to sustainably harvesting vital resources such as water. They also thought long term. Some of their grandest works have survived for hundreds and even thousands of years, still delivering their services. Some have never been bettered, and will probably often see off our more temporary structures.

I saw this in practice as I drove through the desert of northern Oman in the Arabian Peninsula. With the mercury hitting forty-seven degrees Celsius in the shade, it was the

Ancient Wisdom Is a Shining Light   87

hottest place on the planet that day. But in Al Farfarah, a small village about half an hour from Muscat, the Omani capital, village leader Ali Al Muharbi took me to the cool shade created by forty hectares of tall date palms. I was there to see how those trees were being irrigated in a desert with no rivers. The water, it turned out, came gurgling down a gently sloping channel from the Hajar Mountains near the village, sourced from a tunnel that extended two kilometres into the mountainside, where it tapped hidden stores of underground water.

This miracle in the desert had been delivering twelve litres of cool mountain water every second, day-in and day-out, for centuries, said Al Muharbi, who was in charge of sharing out the precious liquid. As well as the date palms, the water irrigated bananas, fruit trees, winter leafy vegetables, and forage crops for livestock. The system, known locally as a *falaj*, had fifty owners – descendants of the people who first dug the tunnel. Each had inherited water timeshares that ranged from thirty minutes per day to several hours. I watched as farm workers meticulously kept to their schedule, blocking and unblocking the distributor channels to each plot using old rags held down by stones. Little had changed since the old days, Al Muharbi said, though today the farmers time their rights with watches rather than by the sundial that still sat in the square, or the stars at night.

These unfailing pre-Islamic feats of hydraulic engineering, which operate entirely by gravity, remain the only water supply for many Omani villages. 'They may be the most ancient community-run systems for managing water in the world,' said Slim Zekri, a water economist at Sultan Qaboos

University in Muscat, as we hurried from the cool of the date palms back into his air-conditioned car.

Such water channels – elsewhere often known as *qanats* – were dug thousands of years ago throughout the arid Middle East. Originally developed by the Persians, one famously watered the grounds of the British embassy in the heart of Tehran. They capture water from many kilometres inside mountains, and often have vertical shafts to allow tunnel repairs to be carried out far from the surface. They are as grand in their way as the Egyptian Pyramids.

I am a huge fan and have travelled to see them in Syria, where they have kept delivering for villages as modern boreholes have run dry; on the West Bank, where Ahmad Qot, a poor Palestinian goat-herder, lifted a manhole to show me a tunnel that was his only source of water; in Israel, where I explored a fluvial underworld with geologist Zvi Ron, who was mapping abandoned tunnels; and in Cyprus, where I held ropes for Yannis Mitsis, the last *qanat*-builder on the island, as he lowered himself into a tunnel beneath an orchard to do running repairs.

Some people regard *qanats* as historic relics. In some places this is true, but only because boreholes have lowered water tables until the tunnels drain only dry rock. But in Oman the government is reviving them as the best way to water remote orchards and villages. And in the southern Spanish city of Seville, the city authorities are digging new *qanats* modelled on those left behind by Islamic occupants a thousand years ago. They plan to use the water to cool buildings in the new super-hot Mediterranean climate. Ancient wisdom, it seems, sometimes cannot be bettered.

6

Eco-restoration Is Happening

Way back in 1992, I visited the Greek island of Zakynthos to report on how Mediterranean ecosystems were being wrecked by mass tourism. Each year, four-fifths of the sea's loggerhead turtles waded ashore along the seven-kilometre stretch of golden sands to lay their eggs on Laganas beach – also famous for its raucous summer nightlife. The result was carnage. Thousands of egg-filled nests hidden beneath the sand were crushed by beach furniture. When the surviving eggs hatched – usually at night at the height of the tourist season – the tiny turtles scrabbled to the surface and followed their instinct to head for the brightest lights. For millions of years, those had been the reflection of the moon and stars on the water, guiding them to sea and safety. By the 1990s, the baby turtles often made instead for the neon lights of the beach bars. Far from safety, many were pecked to death by hungry seabirds.

The entire turtle population seemed doomed, I told my readers. But seven years later, after growing pressure from conservationists, the Greek government made the beach and its bathing waters into a marine park. It banned boats, which often collided with incoming turtles, set aside parts of the beach exclusively for turtle nesting, and prohibited people and lights at night. It worked. A quarter of a century on, the hatchlings that made it back to sea in the 1990s are returning as adults to lay their own eggs. They are finding the beach a much more congenial place. Conservationists reckon the number of eggs hatching in the sands has doubled, and the survival rate for new hatchlings is hugely improved.

This is a small success story, perhaps, but one being repeated widely. According to a global survey in 2025, sea turtle populations are recovering in the majority of places where they still nest around the world. Thanks to growing awareness of their plight, both hunting and coastal developments on their beaches have been reduced, though entanglement in fishing gear at sea remains a threat.

In the Anthropocene, extinctions continue to take their toll on biodiversity. Lonesome George, the last Pinta giant tortoise on the Galápagos Islands, died in 2012. Other recent losses include the Western black rhino, the Caribbean monk seal, the Chinese paddlefish and the Caspian tiger. But the rate of known losses has been declining, and there are ever more victories for species restorations on land and at sea.

The Spix's macaw, declared extinct in the wild in 2019, is now back thanks to reintroductions. The recovery of whale populations since an international moratorium on whaling was introduced in 1986 is well-documented. Today, 'there are more whales in the sea than any living human being has ever seen,' says naturalist Sir David Attenborough. Even less obvious candidates for public sympathy are often on the rebound. In northern Australia, the authorities introduced a ban on hunting saltwater crocodiles, the largest reptiles on the planet, in 1971. Today, the number of these fearsome man-eaters in the wild has risen from 3,000 to 100,000, and it is estimated that they consume nine times more prey per square kilometre (not usually humans) than they did fifty years ago.

Or take tigers. They are big and aggressive and need lots of space. You might think this is not a good combination in a densely populated world. Especially perhaps in India, the world's most populous nation, where Bengal tigers kill around fifty people every year. Yet, India is home to three-quarters of the world's wild tigers; protected from poaching, their number has doubled to more than 3,600 in just over a decade. Their heartlands are designated tiger reserves rich in prey, but they roam much more widely and are regularly reported in surrounding rural areas covering roughly half the size of Britain. This is land that they share with some sixty million people.

There is a dark side to this species restoration, as some of the tiger reserves have been taken from tribal communities. That was both unjust and counter-productive, since tigers loved these places already, precisely because the habitat was protected by tribal communities. This land should be given back. Still, the wider picture is a remarkable success story, built in large part on consent from locals who gain from development projects that help them coexist with the tigers. 'The common belief is that human densities preclude an increase in tiger populations,' says Yadvendradev Jhala of the Wildlife Institute of India, author of the most recent tiger count. 'What the research shows is that it's not the human density, but the attitude of people, which matters more.'

While India's tiger reserves are largely state endeavours, in the U.S. private philanthropy plays a much bigger role in conservation of the country's big beasts. In Montana, America's third emptiest state, rich environmentalists are buying up unprofitable cattle ranches, tearing down fences and restocking with buffalo. Once, some seventy million of these great beasts snorted and grazed their way across the Great Plains, all the way from Alaska to the Gulf of Mexico. They were central to the Indigenous way of life, providing meat, hides, bones and horn for food and clothing, shelter and tools for native Americans. European colonists ended all that, exterminating the bison in large part to snuff out native livelihoods. At the lowest point a century ago, there were barely 500 buffalo left, often on First Nation reservations.

But thanks to breeding programmes and reintroductions, there may now be half a million all told. Admittedly, most are behind fences in commercial herds, such as those supplying media magnate Ted Turner's bison burger chain. Still, around 20,000 roam wild. That should be just the start of the restocking, says Sean Gerrity, a former Silicon Valley entrepreneur who founded Montana's biggest land buyer, the American Prairie Foundation. It is well on the way to assembling 200,000 hectares (think twenty-two Manhattans) to turn into 'America's Serengeti', where grazing buffalo can remake prairie ecosystems fit for the return of other denizens of the grasslands, including prairie dogs, cougars, black-footed ferrets and pronghorns.

In 2010, I went to see. It was quickly clear that not everyone out on the plains of Montana appreciated the West Coast rewilders taking their land, or indeed conservationists in general. I saw plenty of bumper stickers declaring SIERRA CLUB SUCKS. 'Most locals hate them. They just buy land. They don't care about the ranchers,' said Don Youngbauer, a near neighbour with his own 10,000-hectare cattle ranch. He didn't share this antipathy, however. He also ran the local Rosebud Conservation District, a long-standing elected body aimed at improving the environment for landowners as well as nature. He thinks the eco-minded newcomers have a point.

'We have a devastated ecology here,' Youngbauer told me as we picnicked on a hilltop, watching the sun go down over his pastures. 'When it was grazed by buffalo, this land was much more productive. The grass grew up to a horse's belly. We have to find a way of recovery.' He has been making his own contribution by changing the way his cattle graze. Instead of letting them wander, he has introduced 'high-intensity, short-duration' grazing that mimics buffalo ways. 'It's a start,' he said.

Call of the Wild

In Europe too, rewilding is gaining momentum. Unlike their American cousins, European bison were hunted to extinction in the wild by the 1920s. Their genes were kept alive only by captive breeding, but today conservation groups such as WWF are restoring wild herds through reintroductions from captive herds. The largest population, of around a thousand animals, lives in the ancient Białowieża Forest on the border between Poland and Belarus.

Meanwhile, other big beasts of Europe are gaining ground thanks to eco-restoration projects, including some iconic carnivores. Twenty years ago, there were fewer than a hundred

Iberian lynx left, skulking in two tiny populations in southern Spain. Many conservationists saw them as doomed to extinction. Now, thanks to improved protection of their habitat, they number more than 2,000. Their close cousin, the Eurasian lynx, is also returning to central and western Europe, where it was effectively extinct half a century ago.

Brown bears are doing even better. Spreading west from Russia and Romania's Carpathian Mountains, there are now more than 25,000 roaming across Europe, a fifth more than a decade ago. Wolverines and jackals are also on the march. Likewise, wolf numbers have more than doubled in the EU. While wolves, as we saw earlier, have regained ground largely on their own, most of these other recoveries did not just happen. The abandonment of farmland has been accompanied by more deliberate conservation actions such as hunting bans and dedicated habitat enlargement, often by linking up shrinking enclaves to allow interbreeding between previously isolated small populations. There have also been reintroductions, both of the animals themselves and of their favourite prey. In the case of Iberian lynx, more rabbits did the trick.

Inevitably, there has been a backlash, especially against the return of hunting animals. Spanish farmers complain that the resurgent lynx attack their sheep, and have successfully opposed some local reintroductions. Following a furore when a brown bear killed a man walking in the woods, the Slovak government has approved quotas allowing hunters to shoot up to 350 bears each year. In 2024, EU officials downgraded the wolf's protected status, allowing countries to green-light hunting. Conservationists and animal rights groups have complained. 'If we expect countries like India or Indonesia to protect their tigers, and Africans to protect their lions and elephants, then we . . . Europeans should be willing to tolerate some wolves,' Laurent Schley, head of wildlife for the Luxembourg government, told the BBC. Be that as it may, the loosening of protection is a sign of a growing conservation success story across much of Europe. The big beasts are back.

In Britain, rewilding is also all the rage in the Scottish Highlands. Elk, wolves, bears and lynx no longer lurk in the glens, and the English Channel makes natural returns from mainland Europe unlikely. But a growing number of local landowners are intent on bringing native vegetation back to replace the denuded grouse moors and sheep estates. Some of them hope one day to reintroduce lost carnivores too.

Up on the north-west coast, around the small fishing port of Lochinver, crofters and local activists have bought part of the giant Assynt estate from the Vestey family, a meat-producing and retailing dynasty that made its fortune a century ago by turning South American grasslands and rainforests into cattle ranches. The new owners are planting native trees and restoring moorland and coastal ecosystems, while a wider landscape partnership is connecting up restored fragments of natural ecosystems across 80,000 hectares of the area. It is 'one of the largest landscape restoration projects in Europe,' woodland manager Elaine MacAskill told me.

The original Assynt purchase, completed in the early 1990s, turned out to be a watershed moment for land reform in Scotland. Many other communities and conservationists have since raised funds to buy out their absentee landlords with the same ecological intent. In 2025, Scottish Wildlife

Trust bought the Inverbroom shooting estate to bring back Atlantic rainforest. In the Scottish Borders, conservationists from Edinburgh are turning an entire valley of former cattle farms into the Carrifran Wildwood, a replica of what they believe stood there 6,000 years ago. With approaching a million trees planted, they now want to stand back and let the natural ecosystem develop in its own way.

For my final adventure in Scottish rewilding, I met the new laird of a restored pine forest north of Inverness. Dutch flat-pack furniture heir Paul Lister wants to bring back the old fauna, as well as the trees. Maybe lynx first. But, he told me as we dined in the grand lodge at the heart of his Alladale Wilderness Reserve, 'if we could introduce a couple of packs of wolves, and maybe a dozen brown bears, that would be my dream. I'd die happy.' For now, however, the only roars you hear echoing through this would-be wilderness are RAF Tornado fighters flying low on training flights. For one terrifying moment, while climbing the peat moor above the lodge, I found myself looking down on two of them swooping up the glen. The return of the long-lost creatures may have a way to go here, but their habitat awaits.

Regreening Drylands

Sometimes rewilding is happening on a grand scale, but it is also going on, quietly but surely, in places you might barely notice. One of my favourites is the restoration of hedgerows. The farming landscapes of Britain and Ireland are still more densely packed with hedges than anywhere in the world. The hedges in England alone would wrap around the world ten times. That's only half the number we had a century ago, but the figure is starting to creep up again as people go out planting at the weekends.

Hedges matter. They are home to rich ecosystems of lichens and mosses, mice and voles, insects and nesting birds. Most in Britain are more than 250 years old, and the older they get, the more species they harbour. One study found more than 2,000 species of plants, mammals, birds, reptiles and more in an eighty-five-metre stretch. No wonder ecologists call them wildlife superhighways, keeping nature going among the crop monocultures.

The important lesson here is that conservationists should not ignore still-productive farmland: it is possible (and beneficial) for nature and human-shaped environments to coexist. In many tropical countries – even those still losing natural forests – there are ever more trees on farmland. They are planted for timber and fruit crops, or put among fields to stabilize soils, hold on to rainfall and nutrients, and shade crops. In Latin America, tree cover on cropland is increasing across Brazil, Argentina and Mexico. In Africa, at least 29 per cent of tree cover is 'outside areas previously classified as forest,' says Florian Reiner, a remote-sensing analyst at the University of Copenhagen. In the Sahel region on the edge of the Sahara, farm trees make up the majority of tree cover.

In Niger in particular, many farmers have changed the way they farm since the devastating droughts of the 1970s and 1980s. They have found that nurturing the natural regeneration of trees in their fields improves their crops. Dutch geographer Chris Reij was the first outsider to recognize the scale of what was going on after returning to the country for the first time in twenty years. 'I drove 800 kilometres east from the capital, Niamey, and I thought: bloody hell, there are trees everywhere. I had never seen such densities on cultivated land. You couldn't see villages because they were hidden behind trees,' he told me. 'It was obvious they took a lot of pride in their achievements. Every tree had been pruned so it would develop a trunk.' He reckons there are now some 200 million more trees across the previously almost treeless desert fringes in the south of the country. Away from the Sahel, Reij has documented farmers in Kenya and Malawi on the same journey. The phenomenon is probably much more widespread than that, but counting trees on farms is still not something that environmentalists do much.

Combined with the outright abandonment of farmland in many developed nations, something important is happening here. An increasing proportion of the world's trees are not in forests, but in human landscapes. Some may despair at this, but I am enthused by it.

Rewilding Wetlands

Ecological restoration isn't just confined to dry land. The U.S., the first home of modern super-dams for generating hydroelectricity, is also leading the way in tearing them down to restore fluvial ecosystems and, particularly on the West Coast, allow the return of once-abundant salmon. The biggest projects completed to date have been on the River Klamath, California's second longest river, where four dams have been removed, and the Elwha River in Washington state, where the sixty-four-metre-high Glines Canyon Dam is gone, restoring salmon runs into the Olympic National Park that had been lost for a century. These were hard-nosed decisions. In both cases, economists had concluded that the salmon fisheries and other recreational activities on the liberated rivers were worth more than the hydroelectricity from the demolished dams.

Something similar has been happening on the East Coast. New England has tens of thousands of small dams on hundreds of rivers, big and small. Most of them were raised in the nineteenth century to power textile mills, paper mills and sawmills. With these industries long since gone, the dams served no purpose and are belatedly being removed too. Some 800 to date, with potentially some 30,000 more to go.

Europe is also removing engineering to restore free-flowing rivers and bring back wetlands. The Netherlands is almost entirely made up of lowlands created over the centuries by draining the shallow coastal waters of the North Sea and the giant delta of the River Rhine. For the Dutch, dykes that hold out water and create drained areas of land, known as polders, have long been an emblem of national survival. But now, in a remarkable twenty-first-century about-turn, they have begun breaking the dykes and bringing back the water – the better to protect the land.

All along the shore of the North Sea, coastal walls raised to protect low-lying pastures are being removed. The aim is not to give the land back to the ocean, but to help the land resist rising tides. Flooding low-lying areas will create room for the natural regeneration of salt marshes. These wetlands accumulate vegetation and sediment, raising the land surface and acting as a buffer against the ocean by absorbing tides and waves and counteracting rising sea levels.

I visited one such project north of the Friesland town of Leeuwarden, where salt-marsh species such as sea plantain, glasswort and seablite raised the land surface by seven centimetres in the first decade. There were wider wildlife benefits too, said my guide, Chris Bakker of the restoration NGO It Fryske Gea. These expanding muddy intertidal areas, where the sea comes and goes, are rich in worms and other critters that attract tens of millions of migrating birds each summer. They are also a vital spawning ground for herring and other North Sea fish.

Inland too, Europeans are undoing centuries of engineering of the country's rivers. Back in the winter of 1991, I visited activists braving winter snows to occupy a site where engineers were going to build another giant dam on the River Loire, France's longest river. It had been a long winter, but the bulldozers had been kept out and, by chance, on my last day, courts handed down a victory, preventing construction.

Since then, on the Loire and other rivers across Europe, dam-building has come to a halt, and we are now in an era of tearing down dams, weirs, sluices and other barricades to free their flow. In 2024, more than 500 barriers were taken down according to an annual report on progress. Many are obsolete, serving no known function any more. But there is a way to go. There are upwards of a million such constructions across Europe's rivers – more than one for every mile of waterway. But in many places, salmon can return to their spawning grounds, eels can wiggle upstream, rivers can reconnect to their floodplains, and fluvial ecosystems can revive.

The Dutch are among the most ambitious fluvial restorers. Their flagship project has reflooded one of the country's largest and most famous polders. Dykes were erected at Noordwaard in the heart of the Rhine delta back in 1421, when a catastrophic flood on St Elizabeth's Day drowned thousands of people. Soon, the Netherlands was on a national mission to tame its rivers. Then, in 1995 the dykes at Noordwaard and many other places failed as the Rhine burst across the land. A quarter of a million people had to be evacuated from their homes. The cry went up as it had in 1421: 'Never again.' But this time the solution was different: they decided to tear down the dykes and give rivers more room by restoring their lost floodplains.

Today, Noordwaard is no longer protected from the river. With its dykes removed and farmhouses demolished, the Rhine is free to flood across a huge new wetland. The water is enjoyed by thousands of tourists each year. And hopefully it also provides a new sense of security: one that comes not from being reliant on dykes, but instead from the resilience of a rewilded river.

Rewetting former wetlands to ward off natural disasters has taken off in Russia too. There, its purpose is to prevent forest fires. Again, a big disaster forced the change. In 2010, huge fires broke out in wide areas of countryside around Moscow. The flames spread underground into dried-out peat bogs, which burned for months, even smouldering through the winter beneath a thick layer of snow. The peaty smoke from the underground fires travelled for tens of kilometres, engulfing the streets of the Russian capital for weeks on end. In the Kremlin they were not happy. Soon after, Vladimir

Putin launched a joint project with German wetland ecologists to prevent fires by blocking farm drains to rewet dozens of the dried-out bogs.

Prior to Putin's invasion of Ukraine, I visited several of these projects with Russian and German experts. Most impressive was the revival of the Meschera bog, east of Moscow. In Soviet times, over 100 square kilometres of boggy forest there had been stripped of its trees, drained and scraped of peat to provide fuel for the nearby Shatura power station. That desecration turned the whole area into a tinderbox by 2010. But now the wasteland has been restored to its former watery and biodiverse glory and designated a national park.

'We are restoring nature, and making a place where people want to come,' said park chief Zoya Drozdova, as we drove down a muddy track that had for many decades been a railway bringing in thousands of miners and taking out the peat. Leaving the track, we took a long walk, exploring the restored creeks, listening to the abundant birdlife, spotting some of the hundreds of species of lichens that have moved back in, and marvelling at the huge beaver lodges erected from nothing in just a few years. Black alder trees grew in profusion, their roots luxuriating in the water. 'My staff call them Russian mangroves,' Drozdova said. Probably Putin doesn't care too much about all that nature, but the natural recovery has ensured that fires have never again wreathed the Kremlin in smoke.

The move towards ecological restoration of wetlands is gaining ground in many developing nations too, especially in countries where, as in Putin's Russia, there are immediate gains for local communities in better harvests of natural resources or protection from the elements. There is a growing movement for restoring the dense thickets of mangroves that once sank their roots into the muddy waters of tropical coastlines. Historically, mangroves have protected many shores from waves, tides, currents, storms and coastal erosion. They have even lessened the impact of the most devastating threat of all – tsunamis. Sadly, in recent decades they have been uprooted from thousands of kilometres of coast across South East Asia, mostly to make room for shrimp ponds. Shrimp harvests can be very profitable for coastal communities. The problem is that the loss of the tangled foliage and sediment-trapping roots of mangroves has left many communities undefended against the ocean. Not least when the Indian Ocean tsunami hit in 2004.

Indonesian fisherman Hajamuddin was at sea that day. He was one of the lucky ones, as the sea turned out to be the safest place to be. Hajamuddin survived, but almost everyone at home in his village, Gle Jong, on the northern tip of the Indonesian island of Sumatra, drowned as the colossal wave swept across a coastline that had recently lost its shield of mangroves. The handful of survivors had rushed up the steps to the village's only high point – ironically, its cemetery. 'My family was all gone,' Hajamuddin told me when we met in the village a decade later. By now, the community was reviving and there had been an influx of newcomers. Hajamuddin had a new wife, and the residents new and old had joined in planting 70,000 mangroves along the shore to help protect against the next tsunami.

'The mangroves absorb wave energy,' said my guide, local environmental scientist Agus Halim of Syiah Kuala University, whose wife and two children had also died in the disaster. Villages that retained their mangroves fared much better during the tsunami. A German study concluded that their removal had cost 10,000 lives, Halim told me as we lunched with his new wife and watched their son play in the sand.

All along the northern and western shores of Sumatra, which felt the full force of the tsunami, I saw similar efforts at restoration. Around two million seedlings had been planted in front of about seventy villages. Not all took root, but many were growing well, and villagers found that harvests of crabs, cockles and other marine life among their roots often compensated for the lost shrimps.

Indonesia is a nation of islands – more than 17,000 in all, 6,000 of them inhabited. That means a lot of shoreline, and a huge risk to inhabitants from the vagaries of the ocean when their mangroves have been removed. On the low-lying north coast of Java, communities have lost large areas of land already as the sea creeps inland. One morning, I drove towards the village of Timbulsloko, not far from the city of Semarang. On my map, the village appeared to be a regular coastal community, reached by a road from inland. The reality was different. For the last five kilometres, the road was flanked on either side by open water where there had once been rice fields. Houses sat half-submerged and a cemetery had been washed away. At the end of what was now a narrow causeway stood the village hall, raised on stilts above the water. Inside, villagers gathered to tell me how they were trying to reclaim their land and protect their surviving homes by restoring their lost mangroves.

Planting mangrove seedlings hadn't worked well here, they said. The tides and currents had washed them away. So, they erected brushwood fences in the shallows to create some still water. 'The fences act like mangrove roots, capturing sediment and acting as a barrier against the waves. New mangroves can now grow in the sediment,' said Mat Sairi, the chair of the village group installing the barriers. In fact, they didn't even have to plant the mangroves, because seeds floating in the water settle and germinate in the silt. 'Nature does the restoration,' he said.

The mangroves are coming back so well in Timbulsloko and neighbouring villages that the government in Jakarta is copying the initiative elsewhere in Java and other islands besieged by the ocean. Dutch advisers on the project call it 'building with nature' and want to try it across South East Asia in places where planting has failed. Just as salt marshes are all the rage in the Netherlands to shore up the coastline, so mangroves are doing the business here.

Let's Not Micromanage

This has been a somewhat patchy survey of the emerging patterns of ecological restoration, through both re-establishing wilderness and rearranging human landscapes to accommodate more nature. In many ways, the patchiness – the diversity – is the point. Nature is a jumble, sometimes made up of established ecosystems but often comprising novel arrangements in a state of flux. Conservationists intent on finding and sustaining their visions of a pristine environment may be confounded. For me, this is the essence of how nature will be in the Anthropocene. We must go with the flow and be prepared for our efforts to work out differently from how we expect – to allow natural forces to do some of the work for us.

The UN has declared the 2020s a decade for ecosystem restoration. It is an important moment. It signals that the world is no longer content to stem the loss of nature; we are now going on the offensive to reclaim some of what has been lost. I have one concern. It would be a shame if this grand project for the Anthropocene was interpreted as a rallying cry to rewind the clock ecologically, to try to recreate in detail what we had before. That won't work. Too much has changed. Eco-restoration these days is generally most successful when it resists the temptation to micromanage the process, and instead creates room for nature to revive in its own way.

Ecosystems are constantly in a state of flux, with their own internal cycles and dynamic trajectories, and are often adapting to outside forces, not least a changing climate. There is no perfected state, no finely tuned balance. Instead, 'whenever we seek to find constancy, we discover change,' wrote Daniel Botkin, an ecologist at the University of California, Santa Barbara, in his book *Discordant Harmonies*. That is nature's strength. We should seek to restore that dynamism, not tame it.

Two well-documented examples illustrate the different approaches and outcomes. When Dutch rewilders attempted to restore nature to the Oostvaardersplassen polder, they had a grand plan. They planted all the vegetation they wanted and stocked it with grazing animals that most closely resembled the extinct species they thought had once lived thereabouts, including Konik horses, Heck cattle and red deer. Things went well to start with. But too well, it turned out. Animal numbers soon exceeded the pastures' capacity, and most of the animals starved to death, forcing an end to the experiment. So much for grand plans.

But when the splendidly named Isabella Tree decided to rewild her 1,400-hectare farm at Knepp Castle in southern England, she opted for something more modest. She avoided planting and instead waited to see what happened when she walked away. Thorny scrub grew up, she found. It wasn't immediately pretty, but birds and winds were dispersing seeds from elsewhere, so nature was at work. 'At the end what you'll get is an open woodscape [with] the next generation of trees coming up.' Her book, *Wilding*, became a bestseller and a textbook for others to follow.

The lesson is that we should avoid setting ourselves up as the arbiters of the right sort of nature for any given place. We are almost bound to be wrong. Instead, we should aim to unleash nature's dynamism, especially when a changing climate means species and entire ecosystems will need to be on the move to keep up.

This is not a manifesto for laissez-faire. Of course, we should protect and enhance places we regard as special – for our own good as much as nature's. And, of course, we should undo many of our most egregious works that impede the forces of nature: by removing dams on rivers and drains that empty wetlands; by retreating from farmland where we can; by removing fences from grasslands; by reducing pollution of the atmosphere, our waterways and oceans; and above all by reining in climate change. But then, as often as possible, we should step back and see what happens. Nature knows best.

7

The Miracle of the Commons

The term 'the tragedy of the commons' was first coined in the nineteenth century, but it was made famous in 1968 by the American ecologist Garrett Hardin. He held that sharing the environment doesn't work. Couldn't work. That whenever people have common ownership of a resource, whether forests or fisheries, unfenced grazing pastures or underground water reserves, it is bound to be destroyed, because common ownership is inherently lawless. If nobody is in charge, free riders rule.

The only rational approach for any individual dependent on that resource is to grab what they can while they can, because in the end, the forest will be lost, the pastures will turn to desert, the fisheries will be fished out, or the water will be pumped dry. Everybody suffers the consequences of the resource's overuse, but in the meantime, those who take the most gain the most. 'Freedom in a common brings ruin to all,' he said. The same would apply to global spaces such as the atmosphere and oceans, which we might call global commons. The oceans will become ever more polluted, and the atmosphere will fill with climate-altering chemicals.

Hardin saw the only answer as eliminating the commonality: to privatize the common resource. If it is owned, he argued, it will be protected as a private asset. Hardin's assessment immediately gained traction within the scientific community. The editors of *Science*, the American journal where his work first appeared, say that more than half a century later it remains one of the most cited papers in its vast library. It is routinely – and uncritically – taught to students of environmental studies.

Hardin had other strong opinions. He argued trenchantly against immigration and in favour of eugenics, which led many in his day to call him a racist and quasi-fascist. Such views made him unexpected company for many environmentalists with very different politics. Still, his conception of the tragedy of the commons remains a lodestone of much green thinking. Its widespread acceptance is a major reason why many environmental activists see no way out of our current crisis: why the Anthropocene feels like a one-way trip to disaster. It is used as an argument for why we are bound to fail in efforts to fight climate change or stem the loss of species. After all, how could we privatize the atmosphere, or biodiversity? But does the doom-laden theory hold true in the real world?

Of course, there are places where we can see the dynamics behind the 'tragedy' in action. Anarchy is often destructive. Free riders are a problem anywhere and need to be reined in, whether they are despoilers of the ocean, polluters of the atmosphere or destroyers of rainforests. But the problem arises when the tragedy is seen as somehow inevitable – as a law of nature, or at any rate of human society. Then it becomes disempowering, a prognostication of an environmental Armageddon, and an excuse for taking commonly owned resources into private hands – for privatizing the planet to save it. So is there another narrative about the commons? Thankfully, yes.

Hardin's nemesis was American economist Elinor Ostrom. She had a much more optimistic view of our ability to work together for the common good. In 2009, she won the Nobel Prize in economics for arguing that most local commons are not unmanaged lawless places at all. Communities of users usually manage their commons collectively and with consent. They often do it very well. Common sense prevails. Her examples – meticulously argued in her classic take-down of Hardin, *Governing the Commons* – ranged from Swiss mountain pastures to community forests in Nepal and Indonesian fisheries.

Far from being the answer to environmental problems, privatization is typically less efficient for humans and worse for nature, she suggested. Too often, it is private owners who are the plunderers, all too willing to exhaust the threatened resource, take their profits and move on. If there is a tragedy, she argued, it is a tragedy of the free market.

The evidence keeps stacking up that Ostrom was right, at least for local commons. As we have seen elsewhere in this book, collectively managed forests are often green islands amid a sea of privatized deforestation. This is evident not just in the Amazon, but in many other places. In Mexico, 70 per cent of the surviving forests are on collectively owned lands. One of my favourite examples of a community success story began in 1990, when the Guatemalan government in Central America decided to create the Maya Biosphere Reserve to protect its greatest ecological jewels. It is almost a laboratory test for Hardin's theory about the commons.

The biosphere reserve had been a conservation dream for many years. But when the government's detailed plans were revealed, foreign environmentalists were angry that too much of the reserve's forest lands were to be entrusted to the local people who lived there, for their collective economic use, while not enough was to be put under state protection. The collectively owned forests were bound to fall victim to the tragedy of the commons, they felt.

For largely political reasons, the government rebuffed the critics and stuck with its plan. But in the thirty years since, something surprising happened. A third of the state-protected park areas have been deforested. The park guards turned out to be no match for cattle ranchers funded and armed by Mexican drug traffickers intent on clearing the forest and laundering their ill-gotten gains. Meanwhile, over the fence, the community-run forests have survived intact, immaculately preserved and policed by local custodians.

The forest communities, mostly descendants of the Mayan civilization that dominated the region more than a thousand years ago, have good commercial reasons for protecting their resource. They sell forest products such as mahogany for guitars, nuts, allspice, latex from sapodilla trees that makes a 'natural' chewing gum known as chicle, and palm leaves exported to the U.S. for floral bouquets on Palm Sunday. Many of the products are Forest Stewardship Council-certified and are sold around the world. The desire of the forest communities to protect their common assets has proved superior to the diligence of the park guards. There were some confrontations between the Mayans and the drug barons, but the communities were no nine-to-five state employees: they stood up and won.

'The forest is an economic asset to the people,' I was told by Juan Giron of the Association of Forest Communities, a strong and disciplined collective organization that has its origins in an old union of chicle tappers. Collective land rights 'guarantee access to the forest, and this access leads us to take better care of these resources'.

In modern Africa too, old prejudices against community-led conservation are being upended. Top-down fenced-in conservation, as represented by state-run national parks, is often failing. Instead, it is herders, armed only with mobile phones, who are seeing off poachers in remote places that park rangers rarely venture, alerting their fellows to animal movements by SMS, and accompanying high-rolling tourists who fund their conservation endeavours.

The benefits for wildlife can be huge. In Namibia, community conservancies cover a fifth of the country. Though individually small, they link up to create wildlife corridors through which animals migrate in search of food and water. They have helped triple the country's elephant population to 24,000 in the past twenty years. 'Namibia is probably the most inspiring country. We've seen a whole rural wildlife economy develop around tourism, recreational hunting and harvesting indigenous plants,' says Fred Nelson, an old Africa hand and CEO of Maliasili, a Vermont-based NGO dedicated to bolstering African community conservation.

But it is far from alone. In Botswana, Nelson says, community rangers fit local lions with radio collars that deliver real-time predator alerts to farmers' cellphones. Probably Africa's most famous wildlife area is the Maasai Mara on the border between Kenya and Tanzania, home to the famous 1.5-million-strong wildebeest migration. Here, just a quarter of the ecosystem is managed by twenty-four community conservancies, yet they contain 83 per cent of its large mammals.

Some old-school conservationists remain horrified at this handover of the continent's wildlife. They still see African rural people as poachers and bush-burners, either active enemies of wildlife or too poor and ignorant to take heed of environmental concerns. And they see rural communities' herds of cattle as inimical to wildlife conservation. The truth, says Wanjiku Kinuthia, a senior manager at Maliasili, is that 'community-led models have quietly surpassed fortress conservation in terms of both land area and impact'. Most of Africa's biodiversity 'depends on lands owned and managed by local communities,' says Nelson.

Common Pastures and Common Fisheries

A similar story of success in commonality holds for many more of the world's great grasslands. They cover around 45 per cent of the world's non-glacial land surface, yet less than 5 per cent have any formal state protection. Most are unfenced common pastures, where nomadic herders have time-honoured customs of moving livestock across the land, unimpeded by fences, private tenure or even, when they can get away with it, state boundaries.

It sounds like a tragedy in the making. But this common ownership seems to work both ecologically and for the livestock that graze the pastures. As Lindsey Sloat of the World Resources Institute puts it: 'Many traditions and Indigenous land management practices developed around ecological cycles associated with grazing lands.' And because it works, it persists. If you want to find intact North American prairie grasslands, the best place to go is the reservations of the native American nations. They have 85 per cent of what is left.

I also saw the virtue of traditional methods of common management amid the yak pastures of the vast Ruoergai plateau in China, on the fringes of Tibet. This is no virgin land, says German ecologist Hans Joosten. Even so, 'thousands of years of grazing have created a new landscape that is one of the most impressive open landscapes in the world'. Under traditional methods of common ownership, handed down from generation to generation, the yak herders follow the rains, seeking the good grass for their animals and avoiding the bad. There have been few fences or private landholders in the way to bar their migrations.

But rather than leaving well alone, the Chinese government in the 1990s began to reassign the open collective pastures to private custody by individual herders. They must have imagined that such private holdings would be better managed. The result was the opposite. As the fences went up, the grasses became degraded because herders could no longer move their livestock in their time-honoured manner. The result, as I saw when I visited in 2019, has been the creation of dust bowls in some places, while the other side of the fence is lush ungrazed grass.

The herders I met recognized the problem and had begun quietly removing the fences between their ranches so that they could once again pool their pastures, moving their yaks around according to the dictates of nature rather than land title. Gao Yufang, a Chinese anthropologist now at Yale, who has studied this in detail, told me that local officials had come to see the wisdom of this, and turned a blind eye when the fences came down.

The world has hundreds of millions of pastoralists, and probably another billion people who combine farming with keeping livestock that graze on common pastures. By some estimates, their livestock occupy more than 40 per cent of the planet's land surface – approaching four times more than the area farmers till. They are central to national economies from Mongolia to Morocco, and from Sudan to Senegal. Alpaca, vicuña, llama and guanaco roam the Andes; reindeer still grace the tundra of North America, Scandinavia and Siberia. Despite this, pastoralism's PR has been dreadful. Stories of how herders overgraze their livestock, causing 'desertification', have become a received wisdom of environmentalism. The myths of the common-land tragedies they cause are often told by farmers who want the pastoralists' land. The tragedy of the commons can be a land grabbers' charter.

But look closer and the evidence disappears like a desert mirage. In most places, the animals grazing the grasses and browsing the bush are good news – 'vital for ecosystem health and productivity', as a report from the International Union for Conservation of Nature put it. Far from wrecking the land, the grazing animals mostly conserve biodiversity, hold back the desert, store carbon and prevent erosion. Pastoralists are flexible and able to react quickly to changing circumstances by selling animals or migrating to greener grasses, unencumbered by rules of individual land ownership and unfettered by state boundaries. Their way of life truly demonstrates the virtues of the commons.

I saw this at work again in the Middle East. Mohammed, a Bedouin herder in the Badia, a region of arid badlands in eastern Jordan, where it nudges up to Syria, Iraq and Saudi Arabia, was a man living in two worlds. For much of the year, he and his sheep led a sedentary life in his village. Then each spring, he phoned his friends to discover where the rains had left lush grass, then loaded his flock into a truck and headed for distant pastures.

Mohammed's forefathers, members of the Anizah tribe of the Bedouin, were full-time nomads, riding camels and shepherding their flocks for thousands of kilometres between the rivers Jordan and Euphrates, unencumbered by national boundaries. Today, they are stuck behind heavily policed national boundaries, though they often sell their animals to fellow tribe members on the other side of the border and buy them back later.

In this in-between world, desert tents made of exquisite woollen fabric are patched with old fertilizer bags, while trucks bump across the Badia delivering barrels of water. Shepherds are in constant touch by phone and drive into towns like Safawi, a truck stop on the road to Iraq, to catch up on the latest gossip. All this sounds like a recipe for ecological disaster. The Jordanian government claims that people like Mohammed overgraze the pastures, and wants them and their livestock to settle in villages and end the seasonal migrations. But that would be folly, says Hilary Gilbert of Nottingham University. Government stories of the ecological threat posed by the Bedouin herds present an 'unchallenged conservation narrative [that] has helped perpetuate Bedouin inequality'. Whereas the truth is that under their stewardship the Badia is alive and well. It is the government's development plans that could destroy it. Mohammed and his fellow semi-nomadic shepherds of the Badia may just turn out to be the wise men.

Fishing communities also get a bad rap as environmental villains, when the evidence is growing that they too are often the best managers of their local fisheries and coastal ecosystems – certainly more effective than state control or the invisible hand of the market. One of Ostrom's studies looked at the lobster fishery of Maine in New England, often called America's most lucrative single-species fishery. She found that local management, through 'harbour gangs', was the driving force behind its continued success. It is when local control falters and outsiders move in that the results are disastrous.

I have seen this for myself at another of the world's great fisheries, the nutrient-rich upwelling Atlantic waters off Mauritania in West Africa. They thrived under local stewardship until the region's government started selling licences to Europeans to trawl the nearby waters.

But revivals can happen. Southern California's sea urchin fishery was under severe threat until local divers agreed their own harvesting limits and enforced them independently of outside authorities. The same happened amid the coral reefs of the Gulf of Cortez on the Pacific coast of Mexico, which the French marine conservationist Jacques Cousteau once called 'the world's aquarium'. But in the 1980s, as foreign trawlers moved in, fish stocks plunged and local fishing communities were in sharp decline.

Then a group of fishers in the village of Cabo Pulmo got together with marine biologists to press the government to ban outside fishers and turn the reefs into a protected area where the locals could take back control. For ten years, the villagers agreed to halt their activities altogether to allow the reefs and their fish to recover. And it worked. Today, sharks and groupers and snappers and jack tuna and much else have returned. The villagers are fishing the reefs again, and making an extra living from snorkelling tourists.

These examples show two things: that the creation of even small marine protected areas can have huge ecological benefits, and that community control is usually the key to sustainability. 'Species come back quickly – in three or five or ten years,' says Boris Worm of Dalhousie University in Canada. 'And where this is done, we see immediate economic benefits.'

The story of the crusade to save the reefs of the Gulf of Cortez has galvanized a growing movement around the world. Many marine biologists pinpoint its success as a turning point for marine conservation, and it has become the template for many more community-controlled marine protected areas in coastal regions elsewhere. In Indonesia, for instance, where coral reefs and their fisheries have been under huge pressure, the government is reviving traditional systems of coastal protection known as *adat*, in which communities agree on fishing rules and periodically halt catches altogether to allow fish stocks to recover. They are to become the basic tool for managing the country's fast-expanding network of marine protected areas.

Sacred Groves Still Flourish

However good communities can be at ecologically benign management of their lands and waters, we will always need some places set aside for environmental protection. We often think this is a new idea. Many environmental histories describe the creation of Yellowstone National Park in the 1870s as the birth of conservation. But actually the creation and policing of protected areas has a far longer heritage, with deep roots in most cultures and religions. Across the world, thousands of sacred forests and groves have survived for many centuries and continue to flourish in the modern era.

Preserved by Hindu villages in India and Catholic communities in the hills of Italy, by native tribes of the Americas and African animists, and by Siberian reindeer herders and Aborigines of the Australian bush, their longevity and survival are thanks to collective management according to traditional religious and spiritual beliefs and taboos. Sacred groves are 'the oldest form of habitat protection in human history,' says Piero Zannini, a conservation biologist at the University of Bologna.

Some are revered; others are feared. Either way, they persist; and in a world largely governed by statutory laws, it turns out that this collective cultural conservation is often more effective. Laws are made far away in national capitals, and government park wardens are usually salarymen, whereas committed locals rule from the heart, and they don't knock off when the sun goes down. The lands they protect form a 'shadow conservation network' and are 'becoming ever more important as reservoirs of biodiversity,' says Zannini.

In Japan today, there are vanishingly few ancient lowland forests, other than those in the grounds of Shinto temples, which are estimated to cover more than 100,000 hectares. In Ethiopia, forests once covered most of the northern province of Amhara, which surrounds Lake Tana, the source of the Blue Nile. But in the past century, more than 90 per cent of those forests have been lost. Almost all those that remain are clustered around Ethiopian Orthodox Tewahedo churches. Their protection goes back at least 1,500 years, to the arrival of Christianity in the country.

Across the whole of Ethiopia, there are some 35,000 sacred forests around churches and monasteries, according to Travis Reynolds of the University of Vermont, who has studied them with local scholar Mesfin Sahle of the Institute for Global Environmental Strategies. Thanks to the protection of church parishioners, they have not only survived but also often extended their range in recent times, even as surrounding agricultural areas have turned to dust. In many areas, they are nature's last stand against desertification.

The origins of many sacred groves are lost in history. In British churchyards, there are revered yew trees and nearby holy wells surrounded by patches of protected vegetation, which often pre-date the church by centuries. The same is true across Europe, where many are on church land but date back to pagan times among Celts, Druids, Gauls, Lithuanians, Estonians, Finns and the Welsh. When Fabrizio Frascaroli of the University of Zurich mapped a network of Catholic sacred natural sites across Italy dedicated to the thirteenth-century nature-loving local, St Francis of Assisi, he found that many go back much further. The evergreen oak woodland beside a Franciscan convent at Monteluco in the hills of Umbria, which is famous for its bird life, was founded in the third century BC. It was originally dedicated to the Roman god Jupiter.

In rural Estonia, many villages have their own sacred forests dating back to pagan nature worship. Today, they are often linked to a popular forest-worshipping political movement called Maausk (meaning 'land faith'). During harvest festivals, disciples leave gifts for their ancestors in shrines amid the trees. These sacred forests have long been bastions against outsiders. When the Red Army invaded the small Baltic nation in 1940, villagers sought refuge among them, and a national partisan resistance insurgency known as the Forest Brothers sprang up. It persisted into the 1950s.

Across the world, such spiritually inspired collective protection of nature takes many forms. In Australia, native Aborigines still recognize 'dreamtime' sacred groves everywhere from the rainforests of Queensland to scraps of woodland in the parched interior. Recently, there were estimated to be more than 100,000 sacred natural sites across India, protected by villagers as sources of firewood, water and livestock fodder, as well as for their spiritual and ecological value.

The Chinese province of Yunnan has distinctive protected rainforests on several hundred holy hills that were traditionally held to be inhabited by gods of the Yi and Dai ethnic groups. In neighbouring Tibet, the Adi and Monpa people have since time immemorial revered the forests of the Yarlung Tsangpo Grand Canyon, the world's longest, deepest and steepest canyon. This sacred ecological Shangri-La contains China's largest and most intact forest, with Asia's tallest tree, a native cypress more than a hundred metres tall and believed to be a thousand years old, as well as the world's greatest diversity of large carnivores. They include bears, wolves, wild dogs, and no fewer than eight large cat species, including snow leopards on the summit, lynx and leopards on the slopes, and Bengal tigers in the jungle on the canyon floor.

Africa too is rich in non-Christian conservation traditions. Sacred sites for nature 'were present long before colonization and have continued since,' says Kouadio Raphaël Oura, a geographer at the Alassane Ouattara University of Bouake in Côte d'Ivoire. 'Rooted in local knowledge and protected by traditional authorities', they are places where the dead are venerated, secret meetings and rituals occur, and shamans make contact with the spiritual world.

In South Africa, I visited a wooded grove on the plains of Pondoland near the shore of the Indian Ocean. It was tiny, covering less than a hectare, but cool, dark and moist inside. Locals said it was a sacred place where they grew medicinal plants. Poisonous plants too, some added. After a series of mysterious deaths during a local feud about whether to keep a mining company off their land, that seemed to me quite likely.

Certainly, some African sacred groves have traditionally been centres of sorcery and protected by scary taboos. But times change. In Ghana, a traditional fetish priest back in the 1820s put his spell on an area of woodland inhabited by monkeys revered in two local villages. The sacred place survived in secret through the colonial era and grew in ecological importance as surrounding forests disappeared. Then in the 1970s, with animist beliefs fading, locals reinvented the grove as a tourist destination: the Boabeng Fiema Monkey Sanctuary. It has its own website now, and a guest house.

Sacred groves are clearly vital parts of our environmental heritage. Yet there are few national and no international inventories, and they continue to be overlooked by both governments and environmental groups. One reason for this oversight is that they are different from most state-protected areas. States like to create large national parks and nature reserves, often remote and usually set apart from human communities. Sacred sites are typically much smaller and mixed in among farms and people. They don't have big signs telling you that you are entering them. Many of the species they harbour survive through interaction with humans. They may be cultivated and harvested for medicines or simply find (or maybe prefer) habitat around farms, homes, or temple precincts.

Whatever their origins, the creation and longevity of these places are testament to the power of religion and tradition as a tool for conservation on common lands, and of the importance of maintaining human connections with nature, rather than forcing them apart in the name of conservation. They are not 'other'. They are part of 'us'.

The protection of common lands in their many forms proves that our species can be less innately selfish than Garrett Hardin assumed. That we can all put aside thoughts of personal gain in order to ensure the greater gains for humanity and our planet. And if we can still manage local commons collectively, then surely, we can also find ways to manage the global commons: our beleaguered atmosphere and oceans. For that to happen, the politics, the technology and the collective will must all come together.

Conclusion

In laying out seven reasons why we can have some hope for our future on this planet – of having a Good Anthropocene – I don't want to sound Panglossian. I have spent a lifetime reporting on the perils we face. They are horribly real. But I am hoping – no, I am expecting – that we and our natural world can find ways of prospering through this century and far beyond. I admit it is a hard sell sometimes, even to myself. The outlook has darkened since the inauguration of Donald Trump in 2025, with similar reactionary politics seeping into the mainstream around the world. But it is in the worst of times that we most need to seek salvation. At root, two things give me hope.

One is the ability of nature to regrow, adapt and restore itself. Even amid the ecological wreckage that our modern industrial societies have created, it finds a way – forests sprout anew, endangered species repopulate their former lands, and even the most toxic landscapes become refuges for nature's diversity. As long as nature retains this ability to push back – to do what it has been doing for hundreds of millions of years on planet Earth – then it will go on.

But to aid that regeneration, we must give ecosystems room to do their work. Sometimes we will step back: as food production systems improve, we can look forward to a steep decline in the land taken for growing crops. Sometimes we will leave behind blighted, polluted and even radioactive land, but even those spaces will often be reclaimed in our absence. Sometimes, too, we will engineer nature's return through ever more ambitious projects for ecological restoration and rewilding.

On other occasions, rather than sparing land, we will share it better with other species – something humans have done through most of history. The surviving wisdom of our world's Indigenous custodians will be central to that, I hope. So will new ways, as technical advances help us to green our cities, banish polluting industries and remake agriculture.

So, my second cause for optimism lies in humanity. Not just in our ability to innovate – to come up with technical fixes for environmental threats – but in our ability to change our ways, to rediscover old wisdom, and to imagine the best, then mobilize and act on it. The complex organization of society is what we do best and what distinguishes us as a species. We are far more cooperative than we often imagine. Even the most cut-throat capitalist requires some social order and observed rules of law and finance in which to operate. Now, in the Anthropocene, we need to cooperate as if nature mattered on a shared planet.

There are hopeful signs. Technical innovations are increasingly geared towards reducing our impact on the planet by more efficiently using scarce resources, reducing pollution and, above all, ending our reliance on carbon-based fuels. Taking these actions to rein in the chaotic forces of climate change is the most urgent priority for both our complex societies and nature.

But technical innovations will not be enough. We need social changes too. One is to put a stop to the widening disparities between the rich and poor, and between the powerful and powerless. They threaten social chaos that could undo all our best collective endeavours. Decent lives for all are a precondition for a decent environment for all.

Another is to restore a sense of commonality to manage the shared ecosystems that underpin everything we do. Faced with the enormity of our environmental challenge, I hope and believe we still have it in us to maintain and restore not just local commons – pastures and forests, fisheries and urban air quality – but the global commons, the oceans and atmosphere that sit outside individual ownership and national jurisdiction. That requires both ambition and optimism.

Finally and above all, we need to push back against pessimism, which is my primary purpose in writing this book. Too many of the tenets of environmental thinking get in the way of finding solutions. The myth of the tragedy of the commons is one. Another has been the belief that somehow spiralling population growth will get us in the end, that we have exceeded the planet's natural carrying capacity. History and the facts say different. In our journey from hunter-gatherers through farming to modern industrial and urban societies, our ingenuity and adaptability have constantly brought us back from the brink. So, while such doom-laden nostrums may be helpful in highlighting threats, if we see them as somehow inevitable, unconquerable laws, then they will be our undoing. We may as well head for the hills and have a party till it's all over.

So, can I be an optimist, despite it all? Yes, I believe I can. The worst could still happen, but it doesn't have to. Pessimism is for defeatists. I have faith in humanity's ingenuity and collective will. Our success as a species has been built not on 'survival of the fittest' but on our unique skills at collective endeavour. The human tribe needs to get its act together fast, to find common cause. But I'd place a bet that we *can* have a Good Anthropocene. Though sadly, at my age, I may not be around to pick up my winnings.

Acknowledgements

It is impossible to compile a sensible list of people to acknowledge for a lifetime of reporting. But let me start with the commissioning editors, without whom most of the travelling and much of the research included here would not have been accomplished.

At *Yale Environment 360*, my editor-in-chief, Roger Cohn, commissioned each of the pieces at that site for which I have provided links. At *New Scientist*, my thanks to past editors Bill O'Neill, Jeremy Webb, Alun Anderson, Michael Kenward, Stephanie Pain, Gail Vines, David Concar and Catherine Brahic. The late John Vidal and Bill O'Neill (again) commissioned my travels for the *Guardian*.

My relationship with the forest NGO Fern, where I have been a board member for a number of years, has been invaluable. Thanks especially to Saskia Ozinga and Hannah Mowat and sagacious fellow board member David Kaimowitz.

I also worked with Wetlands International and its longtime CEO Jane Madgwick, culminating in our co-authorship of the book *Water Lands*. For parts of that project included here, my thanks to Tatiana Minayeva in Russia, Xiahong Zhang in China and Nyoman Suryadiputra, Apri Susanto Astra and Yus Rusila Noor in Indonesia. Tom Griffiths of the

Forest Peoples Programme took me to visit the Wapichan in Guyana, where Nicholas Fredericks and Patrick Gomes were great hosts. I am also grateful for the support and insights of Andy White during my association with the Rights and Resources Initiative.

I collaborated with David Venables at the London office of the American Hardwood Export Council and director Petr Krejči on the film project *Forested Future*, for which I visited the Menominee in Wisconsin, where thanks also go to Nels Huse, and saw the recovery of the forests of the Appalachian Mountains in Pennsylvania and North Carolina.

I visited the American Prairie Foundation in Montana in the company of Nicole Divine of the Yellowstone River Conservation Districts Council. Thanks to Slim Zekri in Oman, Zvi Ron in the water tunnels of Israel, ICARDA in Syria, Oxfam on the West Bank and Dick Forslund and the Amadiba Crisis Committee in Pondoland. In Scotland, Paul Lister invited me to Alladale and the estimable Mandy Haggith to Lochinver.

Michael Coe took long flights and even longer bus journeys to meet me in Mato Grosso. Gennady Laptev and Sergey Gaschak invited me into the Chornobyl exclusion zone. Baba Isao guided me around the badlands of Fukushima. A Pulitzer Center travel grant got me to Sarawak and Pondoland. WWF invited me to meet the Gwich'in in Alaska and the fishers of Mauritania. I travelled to the Badia with the help of the Royal Geographical Society.

David Western encouraged and facilitated my interest in community conservation in Kenya, including my visit to Il Ngwesi. Thanks also to Mike Dyer for his bush-flying skills, to

Chris Reij, Fred Nelson and the late Michael Mortimore for being on hand over many years to offer good news stories from Africa, and to Jesse Ausubel for being a fearless contrarian.

Chris Thomas tutored me on the new ecology, Erle Ellis on the deep history of the Anthropocene, Jim Lovelock on Gaia's resilience and the late Jack Caldwell and UN's Joseph Chamie on demographic trends. Chris Goodall first opened my eyes to peak stuff, and Emma Maris to the joys of the Rambunctious Garden. Claude Martin was a brave advocate for new thinking while at WWF and brought me in to write *Treading Lightly*.

Ted Nordhaus, Alex Trembath and colleagues at the Breakthrough Institute invited me to several of their annual dialogues on eco-modernism in Sausalito, California. Mark Sutton invited me to workshops exploring what to do about nitrogen.

Let me finally thank some of those who have helped turn my journalism into book form. Jessica Woollard, my long-time agent, fired my enthusiasm for writing at length, initially at the suggestion of her father, broadcaster William Woollard. Laura Barber has been a fantastic commissioning editor at Granta. For this book, copy editor Jack Alexander offered insights to go with his subbing skills. Thanks also over the years to Amy Caldwell at Beacon Press, Susanna Wadeson at Transworld and Sir Tim Smit at the Eden Project. Your fingerprints too are here.

Notes and Further Reading

This book is not an academic text of the sort that requires full citations and footnotes. But it is, I hope, fully grounded in reality and rigorously sourced. The opinions are my own, but the facts are as truthful as my fact-checkers and I can make them. If you find otherwise, please let me know by email at pearcefred1@hotmail.co.uk.

What follows is a selective list of key written sources that I found valuable, as well as things that you might want to check out further. It also includes links to my previously published reporting, where it might offer further insights and links to sources, or information about places I have visited. I have not usually listed books where the title and author are identified in the text. Nor have I provided links to readily available background material that I reckon you can easily find yourself.

For academic papers, I have generally identified the publication and year, along with the DOI. For more general media content, I have given URLs. Websites can change with time, or disappear altogether, but the links included here were all last accessed in July 2025.

INTRODUCTION

Rockström's most recently available take on breached planetary boundaries was published in *Science Advances* in 2023: DOI: 10.1126/sciadv.adh2458. Sir Martin Rees's pessimism appears in *Our Final Century* (Penguin Books, 2003). Julian Simon's most important work is *The Ultimate Resource* (Princeton University Press, 1981). My first book on climate change was *Turning Up the Heat* (The Bodley Head, 1989).

The opposing world views of Borlaug and Vogt are profiled in Charles Mann's *The Wizard and the Prophet* (Alfred A. Knopf, 2018), and Simon and Ehrlich's wager is the subject of *The Bet* by Paul Sabin (Yale University Press, 2013).

I NATURE IS FINDING A WAY

My past reporting from Sarawak appeared in *Green Warriors* (The Bodley Head, 1991), in *New Scientist* in 1994 at https://www.newscientist.com/article/mg14419534-000, and in 2017 at my other main journalistic outlet, *Yale e360*: https://e360.yale.edu/features/how-protecting-native-forests-cost-a-malaysian-activist-his-life-bill-kayong. Daskalova first published her findings in *Science* in 2023: DOI: 10.1126/science.adf1099. Read more on Europe's abandoned farmland at https://norwegianscitechnews.com/2020/02/abandoned-cropland-helps-make-europe-cooler/.

Nancy Baker is profiled at https://northernwoodlands.org/blog/article/stewardship-nancy-baker, and in my 2024 article for the American Hardwood Export Council in an online series titled *Stewards of the Forest*: https://www.americanhardwood.org/en/news-feed/stewards-of-the-forest-bust-and-boom. The study on the cooling effect of the revived Appalachian forests appeared in

2023 in *Earth's Future*: DOI: 10.1029/2023EF003663. Kurganova wrote on Russia's post-Communist forest resurgence in *Global Change Biology* in 2013: DOI: 10.1111/gcb.12379.

I wrote about my 2016 visits to Chornobyl's exclusion zone here: https://www.newscientist.com/article/2080293, and to Fukushima here: https://e360.yale.edu/features/fukushima_bitter_legacy_of_radiation_trauma_fear, and about Ukraine's wartime rewilding here: https://e360.yale.edu/features/ukraine-war-wilding.

Rayden was writing in 2023 in *Conservation Biology*: DOI: 10.1111/cobi.14163. I interviewed Thomas in 2014 for *New Scientist*: https://www.newscientist.com/article/mg22129510-400, and he later published *Inheritors of the Earth* (Allen Lane, 2017). My well-thumbed copy of Richard Mabey's *Weeds* was published by Profile Books in 2010. Ellis wrote about the virtues of our 'used planet' in 2012 in the *Proceedings of the National Academy of Sciences (PNAS)*: DOI: 10.1073/pnas.1217241110. Read about London's clever pigeons in Roger Highfield's news article 'Pigeons hop on the tube to save their wings', *Daily Telegraph*, 29 September 1995. http://www.iankitching.me.uk/humour/pigeons-tube.html.

2 THE POPULATION BOMB IS BEING DEFUSED

Read about the UN's notorious population awards to Gandhi and Qian at https://www.jstor.org/stable/1973563, which I reported under the headline 'UN defends tactics for population control' (*New Scientist*, 9 August 1984, p. 4). Westing floated the idea of global birth permits during a meeting in Edinburgh in 1987, which I reported in the Feedback column of *New Scientist* (8 October 1987, p. 69).

I reported the Cairo conference under headlines such as 'Women's rights dominate Cairo Plan' and 'Fundamentalists reject choice for women' (*New Scientist*, 17 October 1994, pp. 2 and 4).

The quoted fertility rates and other demographic data in this chapter are all from UN databases or other public sources. My book *Peoplequake* was published in 2011 by Eden Project Books. Why the real global 'replacement' fertility rate is around 2.3 is discussed in *Population Research and Policy Review* (2003) at DOI: 10.1023/B:POPU.0000020882.29684.8e.

My 2002 take on how the baby boom was turning to bust is here: https://www.newscientist.com/article/mg17523525-400. The 2008 words of Singapore's Lee are at https://www.pmo.gov.sg/Newsroom/National-Day-Rally-2008.

I wrote about Germany's demographic implosion for the *Guardian* here: https://www.theguardian.com/world/2010/feb/01/population-crash-fred-pearce. The world passed peak child, a term coined by the late Swedish statistician Hans Rosling, around 2020: https://ourworldindata.org/data-insights/the-world-has-passed-peak-child.

The IMF's report *Aging Is the Real Population Bomb* is at https://www.imf.org/en/Publications/fandd/issues/Series/Analytical-Series/aging-is-the-real-population-bomb-bloom-zucker. For a more positive take, my eyes were opened by Theodore Roszak's *Longevity Revolution* (Berkeley Hills Books, 2001).

Read more about today's new natalism here: https://theconversation.com/russia-is-paying-schoolgirls-to-have-babies-why-is-pronatalism-on-the-rise-around-the-world-258979, and about Putin's ideological war on childlessness here: https://www.dw.com/en/whats-behind-russias-plan-to-ban-child-free-ideology/a-70324064.

3 PEAK STUFF IS ON THE HORIZON

Peak car is deconstructed here: https://www.itf-oecd.org/sites/default/files/docs/dp201213.pdf. London's story, as told by Transport for London in 2020, is at https://content.tfl.gov.uk/travel-in-london-report-13.pdf. And tune in to Gen-Z here: https://www.theguardian.com/money/2021/apr/05/number-of-young-people-with-driving-licence-in-great-britain-at-lowest-on-record.

Belk's famous 1988 paper was published in *Journal of Consumer Research* at https://www.jstor.org/stable/2489522. And the *New York Times* revealed how Americans are eating less at https://www.nytimes.com/2015/07/25/upshot/americans-are-finally-eating-less.html. The Royal Society endorsed Whitmarsh on dematerializing in 2017 in its *Philosophical Transactions A* at DOI: 10.1098/rsta.2016.0376.

The 60 per cent gain in resource efficiency comes from the OECD in 2020: https://www.oecd.org/content/dam/oecd/en/publications/reports/2020/07/improving-resource-efficiency-and-the-circularity-of-economies-for-a-greener-world_1a8b7965/1b38a38f-en.pdf. See here for the collapse of the camera industry: https://petapixel.com/2024/08/22/the-rise-and-crash-of-the-camera-industry-in-one-chart/.

U.K. statistics on materials used per head from the Office for National Statistics are at https://www.ons.gov.uk/economy/environmentalaccounts/articles/ukenvironmentalaccountshowmuchmaterialistheukconsuming/ukenvironmentalaccountshowmuchmaterialistheukconsuming. European stats are at https://www.eea.europa.eu/en/analysis/indicators/europes-material-footprint, and American data are at https://reason.

com/2019/10/09/the-economy-keeps-growing-but-americans-are-using-less-steel-paper-fertilizer-and-energy/.

Hitting the 100 billion tonnes mark is recorded thus: https://www.theguardian.com/environment/2020/jan/22/worlds-consumption-of-materials-hits-record-100bn-tonnes-a-year, and humanity's 'anthropomass' is here: https://www.theguardian.com/environment/2020/dec/09/human-made-materials-now-outweigh-earths-entire-biomass-study. Milo's 2019 take is in *Nature* at DOI: 10.1038/s41586-020-3010-5. China's past and future footprint is tracked in 2022 in *One Earth* at DOI: 10.1016/j.oneear.2021.12.011.

The eco-modernist manifesto is here: https://www.ecomodernism.org/.

I caught up with Rao's work here: https://www.sciencealert.com/this-is-how-much-energy-we-all-need-to-fill-our-needs-and-live-a-modestly-decent-life, and in *Environmental Research Letters* in 2021 at DOI: 10.1088/1748-9326/ac1c27.

4 TECH FIXES CAN WORK

Climate COP outcomes were drubbed by the great and good here: https://www.theguardian.com/environment/2024/nov/15/cop-summits-no-longer-fit-for-purpose-say-leading-climate-policy-experts.

China's stealing a march with green tech is analysed by the estimable *Carbon Brief* here: https://www.carbonbrief.org/analysis-clean-energy-just-put-chinas-co2-emissions-into-reverse-for-first-time/, while we see how solar power stole a march on everyone here: https://www.theecoexperts.co.uk/news/is-renewable-energy-cheaper-than-fossil-fuels#, and here: https://ember-energy.org/latest-insights/global-electricity-review-2025/.

America's pre-Trumpian progress is benchmarked here: https://www.canarymedia.com/articles/clean-energy/chart-96-percent-of-new-us-power-capacity-was-carbon-free-in-2024, and the greening of British boardrooms showed up here: https://www.theguardian.com/environment/2025/feb/24/britain-net-zero-economy-booming-cbi-green-sector-jobs-energy-security, and here https://eciu.net/media/press-releases/2025/uk-net-zero-economy-grows-10-in-a-year-finds-new-report.

Yadav sounded off at the Glasgow COP here: https://www.bbc.co.uk/news/world-asia-india-59286790#. But India's real trajectory is exposed here: https://ieefa.org/articles/surge-indias-renewables-tendering-set-keep-coals-share-below-50-total-installed-capacity#. And China's is here: https://balkangreenenergynews.com/irena-china-has-64-share-in-2024-renewables-growth-half-of-worlds-solar-power-capacity/#.

I explored finding space for renewables here: https://e360.yale.edu/features/solar-land-grabs-agrovoltaics. See also https://www.ciphernews.com/articles/land-is-hard-to-find-for-solar-farms-asias-answer-float-them/. The little prairies of Minnesota are sprouting at https://www.anl.gov/article/insect-populations-flourish-in-the-restored-habitats-of-solar-energy-facilities.

I visited Entasopia in 2015 for *Yale e360*: https://e360.yale.edu/features/african_lights_microgrids_are_bringing_power_to_rural_kenya.

Way back, I told the story of pea-souper smogs and acid rain in *Acid Rain* (Penguin Special, 1987), and related the 1999 horrors of Shenyang in *New Scientist* here: https://www.newscientist.com/article/mg16522222-600.

How China subsequently cleaned up is revealed at https://ourworldindata.org/data-insights/china-has-reduced-sulphur-dioxide-emissions-by-more-than-two-thirds-in-the-last-15-years. Bauer painted the global clean-up picture in *Journal of Advances in Modeling Earth Systems* in 2022 at DOI: 10.1029/2022MS003070. But how far South Asia lags is clear from this: https://www.theguardian.com/world/2024/nov/01/lahore-delhi-choked-smog-pollution-season-india-pakistan.

The shocking toll of lead in petrol was underlined most recently in *Journal of Child Psychology and Psychiatry* in 2024 at DOI: 0.1111/jcpp.14072 and reported at https://www.nbcnews.com/health/health-news/lead-gasoline-tied-millions-excess-mental-health-disorders-study-rcna182881.

I followed the legacy of the man who invented both CFCs and lead in petrol, Thomas Midgley, in this 2017 article: https://www.newscientist.com/article/mg23431290-800. Solomon's 2022 quote is here: https://www.bbc.co.uk/future/article/20220321-what-happened-to-the-worlds-ozone-hole.

The land-saving benefits of the Green Revolution were laid out in *PNAS* in 2013 at DOI: 10.1073/pnas.1208065110. But the downside is here: https://www.anthropocenemagazine.org/2023/06/two-thirds-of-fertilizer-is-lost-to-run-off-this-invention-could-reclaim-it/, and here: https://e360.yale.edu/features/can-the-world-find-solutions-to-the-nitrogen-pollution-crisis.

The IEA's assessment of the energy footprint of AI is reported thus: https://www.iea.org/news/ai-is-set-to-drive-surging-electricity-demand-from-data-centres-while-offering-the-potential-to-transform-how-the-energy-sector-works. Meanwhile, I interviewed Ausubel for *New Scientist* here: https://www.newscientist.com/article/mg18925361-500.

5 ANCIENT WISDOM IS A SHINING LIGHT

I wrote about my 2015 visit to the Wapichan in a report for the Forest Peoples Programme, *Where They Stand*: https://www.forestpeoples.org/fileadmin/uploads/fpp/migration/publication/2015/10/where-they-standweb-spreads.pdf.

Read more about the Madagascar periwinkle at https://www.chelseaphysicgarden.co.uk/the-many-lives-of-a-medicinal-superstar-delving-into-the-global-story-of-catharanthus-roseus/.

The Myers quote opens a chapter in his book *The Primary Source* (Norton, 1992). The figure of 35,000 DRC expulsions is from anthropologist Kai Schmidt-Soltau: https://www.fmreview.org/schmidt-soltau-htm/.

I reported early WWF strife over native expulsions in *Green Warriors* (*op. cit.*), and subsequently in *New Scientist*: https://www.newscientist.com/article/mg17824005-300, where I quoted Martin's warning from a now long-deleted WWF Feature titled 'Imperialism by another name?' The published version of my 2004 report *Treading Lightly* appeared at https://wwf.panda.org/wwf_news/?15290/Treading-Lightly-the-origins-and-evolution-of-WWF, but the text no longer appears to be available online.

The Batwa's recent travails are here: https://www.theguardian.com/global-development/2020/feb/07/armed-ecoguards-funded-by-wwf-beat-up-congo-tribespeople, and here: https://e360.yale.edu/features/batwa-kahuzi-biega-national-park-drc.

WRI's 2014 report, *Securing Rights*, is at https://www.wri.org/initiatives/securing-rights. NASA images capturing the greenness of indigenous reserves in the Amazon can be found at https://earthobservatory.nasa.gov/images/151921/indigenous-communities-protect-the-amazon. I reported on my journey to Mato Grosso here: https://e360.yale.edu/features/amazon-watch-what-happens-when-the-forest-disappears, and in my book *A Trillion Trees* (Granta Books, 2021).

I first wrote about my visit to Il Ngwesi in *The Landgrabbers* (Eden Project Books, 2012), and caught up for *Yale e360* in 2022: https://e360.yale.edu/features/amid-pandemic-tribal-run-conservation-in-africa-proves-resilient.

I reported my encounter with Duquain for the American Hardwood Export Council in an online series titled *Stewards of the Forest* at https://www.americanhardwood.org/en/news-feed/stewards-of-the-forest-tribal-knowledge-meets-ecological-science.

New Scientist reported how cool fires could have prevented hot fires in LA in 2025 here: https://www.newscientist.com/article/2465494. I interviewed Pyne for the same magazine in 2000: https://www.newscientist.com/article/mg16822644-800.

Read about Sámi pasture management in *Forest Ecology and Management* at DOI: 10.1016/j.foreco.2009.07.045, and the Borneo thoughts of Peters at https://e360.yale.edu/features/lessons-learned-from-centuries-of-indigenous-forest-management.

The WRI quote on shifting cultivation appears in its 1997 report *The Diversity and Dynamics of Shifting Cultivation*: http://pdf.wri.org/diversitydynamicscultivation_bw.pdf. WWF's contrary view has been removed from its website but can still be found here:

https://vestibulares.estrategia.com/public/questoes/Forest-fires-the-good541cf513e23/.

Downey is quoted at https://www.sciencedaily.com/releases/2023/11/231128132310.htm#, commenting on his 2023 paper in *Communications Earth & Environment*: DOI: 10.1038/s43247-023-01089-6. I reported on my visit to the Gwich'in in *New Scientist* in 2000: https://www.newscientist.com/article/mg16722502-600. The most recent Porcupine herd census is here: https://www.gov.nt.ca/ecc/sites/ecc/files/resources/fact_sheet_porcupine_caribou_en.pdf. Ellis's words appeared in *PNAS* in 2021 at DOI: 10.1073/pnas.2023483118.

I reported my 2015 Oman water adventure at https://e360.yale.edu/features/oasis_at_risk_omans_ancient_water_channels_are_drying_up. My other *qanat* escapades are summarized in my book *When the Rivers Run Dry* (Granta Books, 2019) and much earlier in *Keepers of the Spring* (Island Press, 2004). *Al Jazeera* reported on the *qanats* of Seville in 2024: https://www.aljazeera.com/features/2024/4/28/how-an-ancient-design-is-cooling-21st-century-streets#.

6 ECO-RESTORATION IS HAPPENING

My original Zakynthos turtle story appeared at https://www.newscientist.com/article/mg13518362-600. And here is the latest global picture: https://cms.archelon.saasify360apps.com/uploads/Wallaceetal2025_Updatedglobalconservationstatusandpriorities-formarineturtles_8786eb7936.pdf.

Indian tiger news from 2025 is here: https://www.bbc.co.uk/news/articles/cly9d4n1rgmo#. Jhala's study in *Science* is at DOI: 10.1126/science.adk4827. Note this earlier proviso from Survival: https://www.survivalinternational.org/news/12991.

Europe's carnivore revival is documented in 2025 here: https://www.theguardian.com/environment/2025/feb/23/europe-carnivores-predators-bears-animals-continent-scientists, with the backlash here: https://www.bbc.co.uk/news/articles/cy4pyw8d-4vzo.

In Scotland, the Carrifran story has been told in *The Carrifran Wildwood Story* by Myrtle and Philip Ashmole (Borders Forest Trust, 2009). I wrote about Assynt and Carrifran in *Return of the Trees*, a 2017 booklet for the European forests NGO Fern: https://www.fern.org/fileadmin/uploads/fern/Documents/Fern%20-%20Return%20of%20the%20Trees.pdf.

Read about the return of hedges in a 2024 report for the UK Centre for Ecology & Hydrology here: https://www.ceh.ac.uk/press/high-tech-aerial-mapping-reveals-englands-hedgerow-landscape, and here: https://www.theguardian.com/environment/2025/apr/17/england-ancient-hedges-wildlife. Robert Wolton tells the story of his super-biodiverse hedge in *British Wildlife* at https://www.north-herts.gov.uk/sites/default/files/2023-09/Footnote%2015%20EH.pdf.

I first interviewed Reij in 2008 (https://www.newscientist.com/article/mg19726491-700) and provided an update on the return of Africa's farm trees in 2023 here: https://e360.yale.edu/features/africa-tree-cover-farmer-managed-natural-regeneration. Read more about Reij's insights here: https://www.reforestationworld.org/voices/chris-reij-senior-fellow-world-resources-institute/. *Nature Communications* published Reiner's paper on the revival of farm trees in Africa in 2023: DOI: 10.1038/s41467-023-37880.

The NGO American Rivers is charting dam demolitions, for instance, here: https://www.americanrivers.org/dam-removal-on-

Notes and Further Reading 145

the-klamath-river/. And read about the New England removals here: https://e360.yale.edu:8443/features/northeast-dam-removals. Europe's story is here: https://damremoval.eu/.

I reported at length about Dutch and Russian wetland restoration, and about putting back mangroves in Java and Sumatra, in my book with Jane Madgwick of Wetlands International, *Water Lands* (William Collins, 2020). See also https://e360.yale.edu/features/a_decade_after_asian_tsunami_new_forests_protect_the_coast on Sumatra and https://e360.yale.edu/features/on-javas-coast-a-natural-approach-to-holding-back-the-waters on Java.

Besides Botkin's book on the new ecology, check out *Uncommon Ground*, edited by William Cronon (Norton, 1996).

7 THE MIRACLE OF THE COMMONS

Hardin's seminal paper, 'Tragedy of the Commons', was published in *Science* in 1968: DOI: 10.1126/science.162.3859.1243. (Read it without a paywall here: https://math.uchicago.edu/~shmuel/Modeling/Hardin,%20Tragedy%20of%20the%20Commons.pdf.) Ostrom's riposte, *Governing the Commons*, was published by Cambridge University Press in 2015. I wrote up the Guatemala story here:https://e360.yale.edu/features/parks-vs-people-in-guatemala-communities-take-best-care-of-the-forest, as well as in *A Trillion Trees (op. cit.)*.

Maliasili's 2024 *Seeding Solutions* report is at https://maliasili.org/resource/seeding-solutions. Read more about lion alerts in Botswana here: https://www.uni-siegen.de/presse/relaunch/en/releases/2017/782540.html, and about the Maasai Mara custodians at https://maraconservancies.org/pdf/A-Thriving-Landscape-An-Impact-Report-MMWCA-compressed-compressed.pdf.

Sloat's WRI 2025 take on grasslands is at https://www.wri.org/insights/grassland-benefits. My visit to the yak pastures of China is recorded in *Water Lands* (*op. cit.*). The IUCN's 2006 *Global Review of the Economics of Pastoralism* is at https://agritrop.cirad.fr/549237/1/document_549237.pdf. My reporting from Jordan's Badia appeared in *The Landgrabbers* (*op. cit.*). Gilbert's critique of mainstream narratives of the Bedouin is in *Biological Conservation*: DOI: 10.1016/j.biocon.2012.12.022.

Read about the harbour gangs of New England at https://www.sandwichmagazine.com/fillings/blog-post-title-four-cty8z-webj5-4frca-lwfrb-y8b9w-r4hkt-k9gx9. I wrote up my journey to West African fisheries in *New Scientist* at https://www.newscientist.com/article/mg17022964-500.

The Gulf of Cortez story arose from my work with the directors of a TV series to coincide with the launch of Prince William's Earthshot Prize, which I related in the book *Earthshot* by Colin Butfield and Jonnie Hughes (John Murray, 2022). Read more about the implications of Ostrom's ideas for the high seas at https://www.mercatus.org/research/policy-briefs/elinor-ostrom-high-seas.

Zannini's 2021 paper on sacred groves is in *Biodiversity and Conservation* at DOI: 10.1007/s10531-021-02296-3. The church forests of Ethiopia are analysed at https://space-solutions.airbus.com/resources/news/various/forestry-and-heritage-the-forest-churches-of-ethiopia/ and in *Global Energy and Conservation* at https://www.iges.or.jp/en/pub/church-forests/en. Frascaroli's 2013 dissertation is at https://www.zora.uzh.ch/id/eprint/93306/1/PhDThesisFabrizioFrascaroli.pdf.

Read about the ecological Shangri-La of China's Grand Canyon at https://woodcentral.com.au/asias-tallest-tree-discovered-in-the-worlds-deepest-canyon/ and at https://e360.yale.edu/features/china-tibet-yarlung-tsangpo-dam-india-water. Check out the Ghanaian monkey sanctuary at https://www.boabengfms.org/.

Index

abortion, 29
acid rain, 3, 59–61
adat, 119
Adi people, 122
Africa, sub-Saharan, 33, 45, 77
African Commission on Human and Peoples' Rights, 74
African megafauna, 77–8, 114
ageing populations, 35–6
Ailanthus trees, 22
air pollution, 37, 57–8, 68
Al Farfarah, 87
Al Muharbi, Ali, 87
Alaska, 85, 92
algae, 64
Alladale Wilderness Reserve, 96
allspice, 112
alpaca, 116
aluminium, 40, 42
Amazon region, 12, 24, 75–6, 85, 111
American Prairie Foundation, 92
Andes, 116
Anizah tribe, 117
Antarctica, 61
antbirds, 71

Anthropocene, 2, 7, 24–6, 46, 69, 90, 106, 110, 126–7, 129, 132
'Anthropomass', 42
Appalachian Mountains, 9, 14–16, 21
Aral Sea, 12
Arctic National Wildlife Refuge, 85–6
Argentina, 35, 97
armadillos, 76
artificial intelligence (AI), 40, 65–8
Association of Forest Communities, 113
Assynt estate, 95
Attenborough, Sir David, 90
Australia, 35, 37
  Aborigines, 120, 122
  saltwater crocodiles, 90–1
Ausubel, Jesse, 68
Azerbaijan, 48

Badia region, 117–18
Baker, Nancy, 14–16
Bakker, Chris, 100
Baltic Sea, 64
Ban Ki-moon, 48
Banaba, 62
bananas, 83, 87

Bangladesh, 30, 40
Baobeng Fiema Monkey Sanctuary, 124
Batwa people, 74
Bauer, Susanne, 58
bears, 9, 13, 19, 94–5, 123
bedbugs, 26
Bedouin, 117–18
beef, 39, 43
bees, 18
Beijing, 52
Belgium, 37
Belk, Russell, 38
Berlin Mandate, 48
Białowieża Forest, 93
Biden, Joe, 3
biodiversity, 10, 13–14, 17, 21, 23–4, 26, 79, 86, 90, 102, 110, 114, 116, 121
birth rates, falling, 30–6
black alder, 102
black-footed ferrets, 92
Black Forest, 59
Black Sea, 20, 64
Blue Nile, 121
Borana Conservancy, 78
Borlaug, Norman, 5, 63
Botkin, Daniel, 106
Botswana, 113
Branco, Rio, 71

147

Brazil, 32, 75, 97
　Pantanal, 12
brownfield sites, 18, 26
buffalo, 77, 92–3
Bulgaria, 12–13, 22, 34

Cabo Pulmo, 119
Cairo population conference, 29–31, 36
Caldwell, Jack, 31
California, 49, 54, 66, 98, 118
Cambodia, 40
Canada, 59, 61
Canvey Island, 18
carbon dioxide, 47, 49, 51, 56, 64, 66, 75
Caribbean monk seals, 90
caribou, 85–6
Carpathian Mountains, 94
carpet beetles, 26
Carrifran Wildwood, 96
Caspian tiger, 90
cassava, 83
Catholic Church, 29
Celts, 121
cement, 65
Center for International Forestry Research, 75
Central Electricity Generating Board (CEGB), 59
childcare, 36
China, 14, 22, 38, 43, 64
　air pollution, 57–8
　electric vehicles, 49
　grasslands, 115–16
　population policy, 28, 30–1, 33–4
　renewable energy, 52, 54
　sacred groves, 122–3
Chinese paddlefish, 90
chlorofluorocarbons (CFCs), 61
Chornobyl, 18–19
Cirata reservoir, 54
Clean Air Act, 57
Climate Change Convention, 48
clothes moths, 26
coal mining, 3
cockroaches, 26
Coe, Michael, 75–6
comb jellyfish, 64
Communism, 12–13
Confederation of British Industry (CBI), 50
COP process, 48–9
copper, 40, 42
coral reefs, 119
corn buntings, 54
Côte d'Ivoire, 123
cotton, 43
cougars, 92
Cousteau, Jacques, 118
crocodiles, 90–1
Cyprus, 88

Dai people, 122
Dalrymple, Sarah, 10
Darwin, Charles, 25
Daskalova, Gergana, 12–13, 22
data centres, 40, 68
Dayak people, 82–3
deforestation, 10–12, 14–16, 21, 64, 73–6, 84, 111–12
DeFries, Ruth, 2, 84
Delhi, 58
Democratic Republic of Congo, 73–4

desertification, 116
desulphurization technology, 60
Didi Chuxing, 38
Dnieper, River, 20
Dogger Bank, 51
Downey, Sean, 84
droughts, 97
Drozdova, Zoya, 102
Druids, 121
Duby, Sam, 56
Dunn, Rob, 25
DuPont, 61
Duquain, McKaylee, 79–80
Dyer, Mike, 78
dykes and polders, 99, 101, 106–7

eco-modernism, 66
eels, 100
Ehrlich, Paul, 4–6
El Salvador, 40
electric vehicles, 49, 60
electricity demand, 41
elephants, 77–8, 95
elk, 20, 95
Ellis, Erle, 24, 86
Elwha River, 99
Endless Mountains, 14
English Channel, 95
Entasopia, 55
environmental Kuznets curve, 39, 43, 58, 65
Estonia, 122
Ethiopia, 121
Euphrates, River, 117
European bison, 93
European Environment Agency, 41

falaj, 87
family planning, 28–31
famines, 4–5, 63
Fawcett, Percy, 75–6

Federovych, Markyevych, 19
Felix, Tessa, 71
fertilizers, 62–6
Figueres, Christiana, 48
fire management, 81–4
Firth of Forth, 50
fish stocks, 53
fishing communities, 118–19
fluvial ecosystems, 98–102
Fonda, Jane, 30
Forest Brothers, 122
Forest Stewardship Council, 112
'fortress conservation', 73
France, 37, 54, 100
Frascaroli, Fabrizio, 121
Fredericks, Nicholas, 70
Fukushima, 18

Galápagos Islands, 25, 90
Gamble, Joseph, 14
Gandhi, Indira, 28
Gao Yufang, 116
Gauls, 121
Germany, 9, 37, 48
 agricultural yields, 62–3
 birth rates, 32–4
 power plant emissions, 59–60
Gerrity, Sean, 92
Ghana, 123–4
Gilbert, Hilary, 117
Giron, Juan, 113
Glasgow COP, 52
glasswort, 100
Gle Jong, 103
Glines Canyon Dam, 99

Gobi Desert, 49
Godwin, William, 5
Gomes, Patrick, 83–4
grasslands and pastures, 114–18
Great Plains, 92
Green Revolution, 5, 63–6
Greenham Common, 17
Greenpeace, 16
groupers, 119
guanaco, 116
guano, 62–3
Gulf of Cortez, 118–19
Gulf of Mexico, 64, 92
Guyana, 70, 83
Gwich'in people, 85–6

Haber–Bosch process, 63–5
Hajar Mountains, 87
Halim, Agus, 104
Hardin, Garrett, 109–11, 124
heathlands, 17
Heck cattle, 107
hedgerows, 96–7
Hellem, Louise, 50
herring, 100
HIV/AIDS, 29
holy wells, 121
human population, 27–36, 62–3
Humboldt, Alexander von, 62
hurricanes, 2
hydroponic farms, 66

ICI, 61
Il Ngwesi, 78
Inca farmers, 62
India, 51, 58, 63, 120
 population policy, 28, 31, 33–4

tiger conservation, 91, 94
Indonesia, 94, 103–5, 111, 119
Industrial Revolution, 4, 11, 17, 41, 67
inequality, 44, 118, 128
infectious diseases, 44
Institute for Global Environmental Strategies, 121
International Energy Agency, 68
International Monetary Fund, 35
International Union for the Conservation of Nature, 116
Inverbroom estate, 96
Iran, 32
Israel, 32, 88
It Fryske Gea, 100
Italy, 32, 37
 sacred groves, 120–2

jack tuna, 119
jackals, 9, 94
jaguars, 71, 76
James, Sarah, 85
Japan, 18, 37, 121
 birth rates, 32–4
 rewilding, 13–14
Java, 54, 104–5
Jevons paradox, 67
Jhala, Yadvendradev, 91
Joosten, Hans, 115
Jordan, River, 117
Jupiter, 122

Kahuzi-Biega National Park, 74
Kaimowitz, David, 75
Kaisa, Nancy, 56
Kakhova hydroelectric dam, 20

Kenya, 55–6, 77–8, 98, 114
Kinuthia, Wanjiku, 114
Klamath, River, 98
Knepp Castle, 107
Konik horses, 107
Kruger National Park, 77
Kubuqi Desert, 52
kudzu vine, 22
Kurganova, Irina, 16
Kuzemko, Anna, 20
Kyoto Protocol, 48

La Rose, Claudine, 71
Laganas beach, 89
Lahore, 58
Laikipia plateau, 78
Lake Superior, 79
Lake Tana, 121
lead, 3, 60
LED light bulbs, 67
Lee Hsien Loong, 32, 36
Leeuwarden, 100
leopards, 77–8, 123
linnets, 54
lions, 77, 95
Lister, Paul, 96
llama, 116
Lochinver, 95
Lohrengel, Mike, 80–1
Loire, River, 100
London, 37, 57
London Underground, 26
Lonesome George, 90
Luxembourg, 95
lynx, 9, 13, 19, 94–6

Maasai, 55, 78
Maasai Mara, 113–14
Maausk, 122
Mabey, Richard, 24
macaques, 18

MacAskill, Elaine, 95
McCall, James, 54
Madagascar periwinkle, 72–3
Mahmud, Abdul Taib, 11
mahogany, 112
Maine lobster fishery, 118
maize, 63
Malawi, 98
Maliasili, 113–14
Malta, 32
Malthus, Robert, 4–5
mangroves, 103–5
Marshall, Lord Walter, 59–60
Martin, Claude, 73–4
Marudi, 11
Mato Grosso, 75–6
Mauritania, 118
Maya Biosphere Reserve, 111–13
Menominee people, 79–81
Merkel, Angela, 48
Meschera bog, 102
Mexico, 32, 54, 97, 111, 118
Mexico City population conference, 27–9, 31
mice, 26
Miliband, Ed, 50
Milo, Ron, 42
Minnesota, 54
Mississippi River, 79
Mitsis, Yannis, 88
Monpa people, 122
Montana, 92–3
Monteluco convent, 122
Montreal Protocol, 61
Moscow, 101–2

Mount Kenya, 78
Mukogodo Forest, 78
Mumbai, 45
Mwangi, Margaret, 55
Myers, Norman, 73

Namibia, 113
Nauru, 62
Nelson, Fred, 113
Nepal, 111
net zero, 49–50
Netherlands, 99–101, 105
New York City, 68
New Zealand, 37
newts, 17
Niger, 97–8
nightingales, 17
Noordward, 101
North Sea, 49, 51, 53, 59, 99–100
Norway, 32, 53, 59

Obama, Barack, 3
Office for National Statistics, 41
Oklahoma, 53
Okoth, Peter, 55
Olympic National Park, 99
Oman, 86–8
Oostvaardersplassen polder, 106–7
optical fibres, 40
Osage Nation, 53
Oshkosh, Chief, 80
Ostrom, Elinor, 111, 118
otters, 71
Oura, Kouadio Raphaël, 123
'outlaw plants', 24
Owino, John, 56
ozone layer, 3, 60–1

Index

Pakistan, 58
palm leaves, 112
Paris Climate Agreement, 1, 48
pastoralism, 116–18
Pennsylvania, 14–16
People's Liberation Army, 28
Pérez de Cuellar, Javier, 28
Peru, 62
Peterborough brick pits, 17
Peters, Charles, 82
pigeons, 26
Pinker, Steven, 5
Pinta giant tortoises, 90
Polos, Kip Ole, 78
Pondoland, 123
prairie dogs, 92
pronghorns, 92
Putin, Vladimir, 101–2
Pyne, Stephen, 82
Pyramids, 88

qanats, 88
Qian Xinzhong, 28
Qot, Ahmad, 88

raccoon dogs, 18
radioactivity, 9, 18–19, 127
Rao, Narasimha, 45–6
rare earths, 40, 43
Rayden, Tim, 21
Reagan, Ronald, 29
recycling, 42, 65–6
red deer, 107
Red Sea, 29
red siskins, 71
Rees, Martin, 2
Reij, Chris, 97–8
reindeer herders, 82, 120
Reiner, Florian, 97–8

renewable energy, 49–56, 66
rewilding, 13–14, 17, 22, 92–108, 127
Reynolds, Travis, 121
rheas, 76
Rhine, River, 99
rice, 63
Rio Earth Summit, 48
robotics, 65
Rockström, Johan, 1, 7, 48, 64
Ron, Zvi, 88
Rosebud Conservation District, 93
Royal Society, 39
Royal Society for the Protection of Birds, 54
rubber, 43
Ruoergai plateau, 115–16
Russia, 9, 13, 16, 94
  birth rates, 32, 34–5
  wildfires, 101–2

sacred groves, 120–4
Sadik, Nafis, 30
Safawi, 117
Sahel region, 97–8
Sahle, Mesfin, 121
St Elizabeth's Day flood, 101
St Francis of Assisi, 122
St Paul's Cathedral, 59–60
Sairi, Mat, 105
salmon, 99–100
salt marshes, 99–100
Sámi, 82
sapodilla trees, 112
Sarawak, 11
Sawariwao, River, 71
Schley, Laurent, 95
Science, 110

Scotland, 95–6
sea levels, rising, 44, 100
sea plantain, 100
sea urchins, 118
seablite, 100
secondary growth, 21
Serengeti National Park, 77
Seville, 88
Seychelles, 62
sharks, 119
Shatura power station, 102
Shaw, Peter, 17
Shenyang, 57–8
shifting cultivation, 83
shrimps, 103–4
Siberia, 12, 116, 120
Silvério, Divino, 76
Simon, Julian, 5–6
Singapore, 31–2, 35
slash-and-burn, 83–4
Sloat, Lindsey, 115
Slovakia, 94
smartphones, 40–1
snappers, 119
Solomon, Susan, 61
South Africa, 123
South Korea, 32, 34
Spain, 32, 37
Spix's macaw, 90
Stalin, Josef, 13
steel, 41–2, 57, 65
sturgeon, 20
Sumatra, 103–4
Susquehanna, River, 14
SUVs, 67
Sweden, 37
swidden farming, 83–4
Swiss mountain pastures, 111

Tamil Nadu, 49
tapirs, 76

Index

Tewahedo churches, 121
Texas, 49
Thames Estuary, 17–18, 50
Thomas, Chris, 23, 25
Thurrock waste lagoons, 18
Tibet, 115, 122
tigers, 90–1, 95, 123
Tilbury power station, 18
Timbulsloko village, 104–5
Time magazine, 27
Tomich, Thomas, 84
tragedy of the commons, 109–10, 116, 128
Tree, Isabella, 107
tree cover, 97–8
Trump, Donald, 1, 3, 42, 48, 50, 86, 126
tsunamis, 103
Turkey, 32
Turner, Ted, 92
turtles, 89–90

Uber, 38
Ukraine, 19–20, 102
ultraviolet radiation, 61
UN Development Programme, 74
UN Sustainable Development Goals, 45, 56
UN World Population Plan, 28

United States
  car use, 37–8
  Corn Belt, 14
  dam removals, 98–9
  decline in consumption, 39, 41–2
  Dust Bowl era, 37
  mental health disorders, 60
  renewable energy, 50, 53
  wildlife conservation, 91–3
  see also individual states
US National Renewable Energy Laboratory, 54
US Supreme Court, 15

Veit, Peter, 75
vicuña, 116
Vietnam, 32
Viter, Stanislav, 20
Vogt, William, 4

Wapichan people, 70–2, 83
washing machines, 42, 45, 55
water management, 87–8
Watt, James, 67
Waukau, Ron, 79
Wayka, Curtis, 81–2
West, Geoffrey, 6
West Bank, 88
Western, David, 77–8

Western black rhino, 90
Westing, Arthur, 29
whales, 90
wheat, 63
Whitmarsh, Lorraine, 39
wild boar, 18–19
wild dogs, 123
wildfires, 2, 81–2, 101–2
Wildlife Institute of India, 91
wolverines, 94
wolves, 9, 13, 19, 94–6, 123
Woods Hole Research Center, 76
World Resources Institute (WRI), 54, 74–5, 84, 115
Worm, Boris, 119
WWF, 73–4, 84, 93

Xingu people, 75–6

Yadav, Bhupender, 52
yak herders, 115
Yarlung Tsangpo Grand Canyon, 122–3
Yellow River, 54
yellowhammers, 54
yew trees, 121
Yi people, 122
Youngbauer, Don, 93

Zakynthos, 89
Zannini, Piero, 120–1
Zekri, Slim, 87